A NATIONAL ASSET

50 YEARS OF THE STRATEGIC AND DEFENCE STUDIES CENTRE

A NATIONAL ASSET

50 YEARS OF THE STRATEGIC AND DEFENCE STUDIES CENTRE

EDITED BY DESMOND BALL
AND ANDREW CARR

Australian
National
University

PRESS

ANU PRESS

Published by ANU Press
The Australian National University
Acton ACT 2601, Australia
Email: anupress@anu.edu.au
This title is also available online at press.anu.edu.au

National Library of Australia Cataloguing-in-Publication entry

Title:	A national asset : 50 years of the Strategic & Defence Studies Centre (SDSC) / editors: Desmond Ball, Andrew Carr.
ISBN:	9781760460563 (paperback) 9781760460570 (ebook)
Subjects:	Australian National University. Strategic and Defence Studies Centre--History. Military research--Australia--History.
Other Creators/Contributors:	Ball, Desmond, 1947- editor. Carr, Andrew, editor.
Dewey Number:	355.070994

Cover design and layout by ANU Press.

Contents

The SDSC team, December 2015

About the Book

This volume commemorates the 50th anniversary of the Strategic and Defence Studies Centre (SDSC). The Centre is Australia's largest body of scholars dedicated to the analysis of the use of armed force in its political context and one of the earliest generation of post-World War II research institutions on strategic affairs.

As a leading international research institution specialising in strategy and defence, SDSC seeks to:

1. provide 'real world'–focused strategic studies that is research-based, research-led and world-class. Our primary expertise within the broad field of strategic studies consists of three related research clusters: Australian defence, military studies, and Asia-Pacific security. Our scholarship in these areas is intended to be recognised internationally and of value to the Australian policy community

2. prepare and educate the next generation of strategic leaders — military, civilian and academic — in Australia and the Asia-Pacific region by providing world-class graduate and undergraduate programs in strategic and defence studies

3. contribute toward a better informed standard of public debate in Australia and the Asia-Pacific region using high-quality outreach and commentary on issues pertaining to our core areas of expertise.

This book contains contributions by the Centre's six successive heads: Dr T.B. Millar (1966–71, 1982–84), Dr Robert O'Neill (1971–82), Professor Desmond Ball (1984–91), Professor Paul Dibb (1991–2003), Professor Hugh White (2004–11) and Dr Brendan Taylor (2011–). It also includes contributions by three of its leading scholars over the half century: Dr Coral Bell, who was present at the creation of the International Institute for Strategic Studies in London in the

1950s, and was a visiting fellow in SDSC from 1990 until her death in 2012; Professor J.D.B. Miller, head of the Department of International Relations at The Australian National University from 1962 to 1987, who, together with Sir John Crawford, then the director of the Research School of Pacific Studies, conceived the idea of the Centre in early 1966; and, finally, Professor David Horner, Australia's foremost military historian, having led several official history projects on peacekeeping and the Australian Security Intelligence Organisation, and who studied at SDSC and joined the Centre full-time in 1990.

These chapters are replete with stories of university politics, internal SDSC activities, cooperation among people with different social and political values, and conflicts between others, as well as the Centre's public achievements. But they also detail the evolution of strategic studies in Australia and the contribution of academia and defence intellectuals to national defence policy.

Contributors

Desmond Ball is a former head of the Strategic and Defence Studies Centre, where he is currently Emeritus Professor.

Coral Bell was a visiting fellow at the Strategic and Defence Studies Centre during the 1990s and 2000s.

Andrew Carr is a research fellow at the Strategic and Defence Studies Centre.

Paul Dibb is a former head of the Strategic and Defence Studies Centre and was appointed Emeritus Professor in 2004.

David Horner is Official Historian and Emeritus Professor of Australian Defence History at the Strategic and Defence Studies Centre.

T.B. Millar was head of the Strategic and Defence Studies Centre from 1966 to 1971 and from 1982 to 1984.

J.D.B. Miller was head of the Department of International Relations at The Australian National University from 1962 to 1987, and a member of the Strategic and Defence Studies Centre Advisory Committee from 1966 to 1987.

Robert O'Neill was head of the Strategic and Defence Studies Centre from 1971 to 1982.

Brendan Taylor is Head of the Strategic and Defence Studies Centre.

Hugh White is a former head of the Strategic and Defence Studies Centre and is currently Professor of Strategic Studies at The Australian National University.

Foreword: From 1966 to a Different Lens on Peacemaking

The year 1966 was by all accounts a troubled one. Tensions boiled over in South Vietnam; civil rights protests escalated in the US; coups erupted in Nigeria and Togo; and the international community struggled to hold a consistent line of action in response to security force killings in Rhodesia. In Australia, the leader of the Labor Party, Arthur Calwell, was shot and injured, and Harold Holt became Prime Minister for a short time before tragedy struck.

Troubled times can generate innovations in peace making, but cooperation, commitment, generosity and further innovation are needed to sustain them. The innovation and labour of peacemaking and peacekeeping at state and supra-state level since 1945 has been much theorised and discussed. In her 1993 report for RAND, for example, Lynn Davis noted that successful interventions for peace need a shared concern about a situation, a desire to put aside vested interests, a commitment to concrete settlements and a recognition of the need for specialist help with elements of solution finding.[1]

Peacemaking and peacekeeping are not just writ by nation states. Nor is the analysis of peacemaking and peacekeeping exhausted by including reference to popular or protest movements like those carefully documented in books like Kyle Harvey's *American Anti-Nuclear Activism, 1975–1990: The Challenge of Peace*.[2] National stances can also rehearse, adopt and adapt institutional voices that run within

1   Davis, Lynn E. *Peace Keeping and Peace Making After the Cold War* (Santa Monica, CA: RAND Corporation, 1993). www.rand.org/pubs/monograph_reports/MR281.html, accessed 5 August 2016.
2   Harvey, K., *American Anti-Nuclear Activism, 1975–1990: The Challenge of Peace* (Basingstoke: Palgrave Macmillan, 2016).

and across state boundaries. Those voices may reflect the input of a relatively small group of people, but the impact of acting on them may operate at global scale.

The Strategic and Defence Studies Centre (SDSC) stands out as one of the few lasting global innovations in peacemaking from 1966. That is easily measured not only in scholarly outputs, but also in government and government agency stances and actions on a wide variety of matters, from nuclear non-proliferation to regional tensions, and from the motivations of those who seek violence as well as peace to the boundaries of domestic security. I suspect, however, that reflection on what has made SDSC successful has understandably taken a back seat to these scholarly and policy outcomes. In cases like this, the view of an outsider can help to throw into focus that which is humbly placed aside in the desire to help others. To my view, advising on peacemaking, peacekeeping, non-proliferation of weapons, threat management, conflict and intervention strategy and non-state fighters requires many of the same skills that play out at state level. Arguably too, the SDSC boasts a record of outcomes that equal or better those of some nation states.

SDSC provides us with an exemplar of what results when we put aside vested disciplinary interests, when we realise that we need to look at complex, dynamic and unstable problems from multiple directions, when we see the power of collaboration across organisations, and when we acknowledge that innovations have to be communicated in multiple ways for multiple audiences. Most of all, SDSC reflects a shared concern in securing a better world.

These norms are in lamentably short supply both within and beyond the academic world. Contemporary funding, policy and scholarly settings tend to drive disciplinary splintering, safe innovations in thought and inward-looking communication. Our times are just as troubled as 1966 and, arguably, the world has more need now for concerted, collective action to ensure that people live with enough safety to access and take advantage of educational, economic, social and cultural opportunities. The stakes are high, and we need to bring the best of ourselves to solving the problems we face.

So while I think it is fitting that we celebrate the scholarly contribution of SDSC to research, policy and action, I also think a celebration of its role as a model in academic practise is 50 years overdue. We need to further our understanding of the forces and eddies in peace, war and tensions, and we need to continue to do that by combining the best of multiple, strong disciplines of theory and empirical work, of the expertise of people inside and outside of the academy, and of different communication styles. Yet there is a further challenge to ensure that academic peacemaking becomes a wider norm. Having reaped the rewards of peace in one place, we have an obligation to make peace in other places. The complex problems that we face domestically and internationally ask no less of us. SDSC has a key role to play in promoting this norm of academic peacemaking and peacekeeping, and I believe that the next 50 years will see it broaden its networks of influence in the pursuit of understanding, policy, and action.

Professor Marnie Hughes-Warrington
Deputy Vice-Chancellor (Academic)
The Australian National University

Preface

The Strategic and Defence Studies Centre (SDSC) was established in 1966. Always a part of the Research School of Pacific Studies (RSPacS)/ Research School of Pacific and Asian Studies (RSPAS) at The Australian National University (ANU), it was for two decades the only academic centre in Australia devoted to research on strategic and defence issues, and is generally acknowledged to be Australia's leading academic centre for research and graduate education in this field.

This volume commemorates the 50th anniversary of SDSC. It contains contributions by the Centre's six successive heads: Dr T.B. Millar (1966–71, 1982–84), Dr Robert O'Neill (1971–82), Professor Desmond Ball (1984–91), Professor Paul Dibb (1991–2003), Professor Hugh White (2004–11) and Dr Brendan Taylor (2011–). It also includes contributions by three of its leading scholars over the half century: Dr Coral Bell, who was present at the creation of the International Institute for Strategic Studies (IISS) in London in the 1950s, and was a visiting fellow in SDSC from 1990 until her death in 2012; Professor J.D.B. Miller, head of the Department of International Relations at The Australian National University from 1962 to 1987, who conceived the idea of the Centre in early 1966, together with Sir John Crawford, then the director of the Research School of Pacific Studies; and, finally, Professor David Horner, Australia's foremost military historian, having led several official history projects on peacekeeping and the Australian Security Intelligence Organisation, and who studied at SDSC and joined the Centre full-time in 1990.

In Chapter 1, Coral Bell describes the formative years of IISS in London, explores the notion of strategic culture in Australia, and places the development of SDSC in these international and domestic contexts. Tom Millar in Chapter 2 and Bruce Miller in Chapter 3 describe the foundation of the Centre; Chapter 4 by Bob O'Neill and Chapter 5 by

Des Ball describe its growth to international repute during the 1970s and 1980s. Chapter 6 by Paul Dibb discusses the Centre's reorientation after the end of the Cold War. In Chapter 7, David Horner reviews of the role of military history in strategic studies and SDSC's role as the single largest centre for the academic study of military history in Australia. In Chapter 8, Hugh White discusses the place of academic strategic and defence studies and, more particularly, the Centre during his leadership between 2004 and 2011. In Chapter 9 Brendan Taylor, the current head of SDSC, concludes the volume. He details the move towards a larger education program and the significant expansion of staff to cover the Centre's contract to provide a course of study at the Australian Command and Staff College in Canberra.

Two of the contributions, those by Tom Millar and Bruce Miller, were originally prepared for a conference held in July–August 1991 to mark the Centre's 25th anniversary. They were published in 1992,[1] and are reprinted here with minor amendments. Chapters 1–6 were published in 2005 in *A National Asset: Essays Commemorating the 40th Anniversary of the Strategic and Defence Studies Centre* and are reprinted here with minor amendments. Chapter 7 was revised by Hugh White to update the record of his full term as head of SDSC.

The Centre is greatly indebted to Bob Cooper and Darren Boyd for their photographic service over nearly four decades, going back to when Bob joined the Photographic Services unit of ANU in 1969. He transferred to the new Coombs Photography unit in 1993, from which he retired in 2005. Darren began at the Photographic Services unit in 1987, and also transferred to Coombs Photography in 1993. They have taken hundreds of photographs of SDSC personnel, conferences, meetings and other activities, many of which are reproduced in this volume. Darren also digitised most of the photographs reproduced herein. Bob O'Neill would like to thank Ross Babbage, David Horner, Jolika Hastings and Sally O'Neill for their comments on drafts of his chapter.

---

1   See Desmond Ball & David Horner (eds), *Strategic Studies in a Changing World: Global, Regional and Australian Perspectives*, Canberra Papers on Strategy and Defence No. 89 (Canberra: Strategic and Defence Studies Centre, ANU, 1992), chpts 3 & 4.

The SDSC 50th Anniversary Conference 'New Directions in Strategic Thinking 2.0' was held on 21 and 22 July 2016. A copy of the keynote address, delivered by Brendan Sargeant, Associate Secretary of the Department of Defence, is included. As is a copy of the conference program for the two days. The authors are grateful for Brendan for contributing his speech to include in the book, and to Professor Marnie Hughes-Warrington, Deputy Vice-Chancellor (Academic), ANU, for providing the foreword to this volume.

Desmond Ball would like to express a special thanks to his wife, Annabel Rossiter, who has shared most of his life with the Centre. She maintains the family photo collection, from which many of this volume's plates were copied. Andrew Carr would like to thank his wife, Katina Curtis, for her assistance with some of the new photos and support during the project.

Desmond Ball and Andrew Carr
Strategic and Defence Studies Centre
The Australian National University, Canberra
August 2016

Acronyms and Abbreviations

ABC	Australian Broadcasting Corporation (known as the Australian Broadcasting Commission prior to 1983)
ACDSS	Australian College of Defence and Strategic Studies
ACSC	Australian Command and Staff College
ACT	Australian Capital Territory
ADC	Australian Defence College
ADF	Australian Defence Force
ADFA	Australian Defence Force Academy
ADSC	Australian Defence Studies Centre
AIIA	Australian Institute of International Affairs
AIPS	Australian Institute of Political Science
ANU	The Australian National University
ANZUS	Australia, New Zealand, United States
ARF	ASEAN Regional Forum
ASEAN	Association of Southeast Asian Nations
ASIO	Australian Security and Intelligence Organisation
ASPI	Australian Strategic Policy Institute
CDF	Chief of the Defence Force
CIA	Central Intelligence Agency (United States)
CMF	Civilian Military Force
CPACS	Centre for Peace and Conflict Studies
CSAAR	Centre for the Study of Australia–Asia Relations
CSCAP	Council for Security Cooperation in the Asia-Pacific
DEST	Department of Education, Science and Training
DFAT	Department of Foreign Affairs and Trade

GRADNAS	Graduate Research and Development Network on Asian Security
GSSD	Graduate Studies in Strategy and Defence
IISS	International Institute for Strategic Studies (United Kingdom)
ISS	Institute for Strategic Studies (United Kingdom)
JIO	Joint Intelligence Organisation
JSSC	Joint Services Staff College
LSE	London School of Economics (United Kingdom)
MDSP	Military and Defence Studies Program
MoU	Memorandum of Understanding
MUP	Melbourne University Press
NATO	North Atlantic Treaty Organization
NSA	National Security Agency (United States)
PRC	Peace Research Centre
RAAF	Royal Australian Air Force
RAN	Royal Australian Navy
RMA	Revolution in Military Affairs
RODC	Regular Officer Development Committee
RSPacS	Research School of Pacific Studies
RSPAS	Research School of Pacific and Asian Studies
RUSI	Royal United Services Institute
SDSC	Strategic and Defence Studies Centre
SEATO	Southeast Asia Treaty Organization
SIGINT	Signals Intelligence
START	Strategic Arms Reduction Treaty
UNSC	United Nations Security Council
USAF	United States Air Force
USI	United Services Institute
UWA	University of Western Australia

List of Plates

The SDSC team, December 2015

Plate 1 Dr T.B. Millar, head of SDSC, 1966–71, 1982–84

Plate 2 Professor Sir John Crawford and Professor Anthony Low, chairmen of the SDSC Advisory Committee, 1966–67 and 1973–75 respectively

Plate 3 Professor J.D.B. Miller, head of the Department of International Relations, 1962–87, and member of the SDSC Advisory Committee, 1966–87

Plate 4 Professor Hedley Bull, joint head of the Department of International Relations and member of the SDSC Advisory Committee, 1967–77

Plate 5 Robert O'Neill, head of SDSC, 1971–82

Plate 6 Colonel J.O. Langtry, executive officer, and Desmond Ball, research fellow, 1975

Plate 7 Joli and Peter Hastings, former research assistant and senior research fellow, May 1986

Plate 8 Robert O'Neill with Gunther Patz, former PhD student, 1992

Plate 9 Colonel J.O. Langtry, executive officer, October 1983

Plate 10 Billie Dalrymple, secretary; Desmond Ball, deputy head; and Ram Subramanian, visiting fellow, Thredbo, 1982

Plate 11 Ray Funnell, Robert O'Neill, Suzanne Funnell, Billie Dalrymple and Mara Moustafine, 1988

Plate 12 Paul Dibb, senior research fellow, 1984

Plate 13 Desmond Ball, head of SDSC, at the entrance to the Pine Gap station, July 1984

1

Strategic Thought and Security Preoccupations in Australia

Coral Bell

This essay was previously published in the 40th anniversary edition. It is reprinted here in its near original format.

The other authors in this volume have provided such authoritative accounts of the processes that led to the foundation, 40 years ago, of the Strategic and Defence Studies Centre (SDSC), and its development since, that there is nothing that I can add on that score. Instead, this essay is devoted to exploring what might be called Australia's strategic culture: the set of intellectual and political assumptions that led to our security anxieties and strategic dilemmas having been perceived and defined as they have been. There will also be some consideration of the outside influences on perceptions in Australia, and of such 'side streams' of thought as have run somewhat counter to the mainstream, but have been represented within SDSC.

Long before Australians constituted themselves as a nation, they developed a strong sense of their potential insecurity, and the inkling of a strategy to cope with that problem. That strategy was what Australia's most powerful and influential prime minister, Robert Menzies, was (much later) to call the cultivation of 'great and powerful friends'. Initially that assumption was so automatic as hardly needing to be defined. Britain was the most powerful of the dominant players in world politics for almost all the 19th century, and Australia was its colony. The ascendancy of the Royal Navy in the sea lines between them provided adequate protection, and Australia's role was merely to provide expeditionary forces (starting with the Sudan War of 1885) for the campaigns in which they might be strategically useful. That tradition has remained the most regular of uses for Australia's armed forces — right up to Iraq and Afghanistan in 2006. Only the identity of the great and powerful friend has changed since 1941.

That change in strategic dependency, from the United Kingdom to the United States, was signalled as a future prospect from early in the 20th century, with the resentment and alarm that the Anglo–Japanese Treaty of 1902 induced in Australia. It was taken by some local observers, quite accurately, to mean that British naval power no longer stretched as far as the Pacific. That was one of the 'subtexts' to the enthusiasm with which the visit of US President Teddy Roosevelt's 'Great White Fleet' was greeted in 1908. It had some embarrassing overtones. Here is a Labor member of parliament, Arthur Griffith, on the occasion:

> A great Anglo-Saxon democracy, Britain's eldest-born daughter, and the wealthiest and most advanced nation in the world, the United States ... has flung down the gauntlet to the Mongol and challenged the naval supremacy of Japan, and by its visit to Australia has given notice to the yellow races that they will have to stop in Asia.[1]

People were not at all coy about racist sentiments in those days.

Despite the prevalence of that type of attitude in 1908, things did not go particularly well between the United States and Australia for most of the next three decades. Australians were disappointed at the delay until 1917 of US entry into World War I. Then, at the postwar talks at Versailles in 1919, Australian Prime Minister William (Billy) Hughes emerged, rather gleefully, as one of the most irritating thorns in US President Woodrow Wilson's side, chiefly over the disposal of the German colony in New Guinea. Hughes obviously enjoyed the role:

---

1   Cited in Coral Bell, *Dependent Ally: A Study in Australian Foreign Policy*, 3rd edn (Sydney: Allen & Unwin, in association with Department of International Relations, ANU, Canberra, 1988), p. 9. See chapters 1 and 2 for a fuller account of this early period.

'Mr. Hughes, am I to understand that if the whole world asks Australia to agree to a mandate in respect of these islands, Australia is prepared still to defy the appeal of the whole civilized world?' 'That's about the size of it, President Wilson' replied Hughes, as he moved his ear-trumpet close to the president.[2]

Later, in the 1920s, there was much Australian irritation at US policy at the Washington Naval Conference (November 1921 – February 1922), and a tendency in Australia to blame the United States for the great Wall Street crash of 1929, and the subsequent depression, which hit Australia particularly hard. Throughout the 1930s, as now, irritation in rural Australia at what seemed to our farmers to be unduly protectionist US policies that kept us out of profitable markets was always a factor. It was therefore not until the approach of war became visible in 1939, and Japan had allied itself with Germany and Italy, that the 'look to America' enthusiasm of 1908 re-emerged in Canberra, and indeed not until the fall of Singapore that it became explicit — it was officially asserted in Prime Minister John Curtin's well-known newspaper article and radio talk at the end of 1941. Neither UK Prime Minister Winston Churchill nor Roosevelt liked those statements, believing the publicity to be alarmist and damaging, but there was obviously no reversing the change given Australia's strategic circumstances at the time.

The six months between Pearl Harbor and the US victory in the naval battle of the Coral Sea in May 1942 seem to me undoubtedly the crucial formative period of Australian strategic thinking ever since, establishing the view that the sea–air gap round Australia was our basic zone of defence of the homeland, and that advanced air and naval capacity was the key to our chances of prevailing there, since the adversary would be almost certain to command more manpower. How that conviction is to be reconciled with the older tradition of an expeditionary force serving with larger allied contingents on distant battlefields, like Iraq and Afghanistan, is still capable of evoking passionate argument between strategists in 2006.

The World War II experience also brought some understanding, at least at the policymaking level, of one of the endemic problems of small and middle powers: that the strategic priorities of one's great

---

2   Bell, *Dependent Ally* (1988), p. 10.

and powerful friends may not be identical with one's own. Britain and the United States were wedded, both before and after Japan struck at Pearl Harbor, to the strategic resolution expressed in the phrase 'Beat Hitler First'. As grand strategy for the overall conduct of the war, that was difficult to quarrel with, but in terms of, for instance, the availability of aircraft to the Pacific theatre as against the Middle East, the consequences were often hard to take in Canberra. Labor's External Affairs minister, H.V. Evatt, spent most of the war making himself disliked in both Washington and London by his constant demands for various kinds of strategic assets or priorities.

The other main strategic dilemma made apparent to Canberra during the later stages of the war in the Pacific arose from Curtin's initial decision to hand over the deployment of Australian troops completely to General Douglas MacArthur. Many people later resented their use in campaigns that cost Australian lives without contributing significantly to the defeat of Japan, which was achieved essentially by US air and naval power, the island-hopping campaign across the North Pacific, and finally the atom bombs. The Curtin Government of 1942 contained no one with experience of command in war, so the initial decision was probably inevitable. Yet, by 1944–45, some modification of that situation could have been sought; or at least some of the 'top brass' in Australia thought so. However, Curtin, by that time, was moving into a sort of twilight of final illness, Ben Chifley was not yet in charge and Frank Forde and Evatt were busy elsewhere. Thus, nothing was done, and some Australians became rather disillusioned not only about MacArthur but about the US connection in general. Perhaps that is why there was a half turn away from the United States and back towards Britain in the immediate postwar period, although Evatt continued to work (hard but unsuccessfully) at extracting a security treaty from Washington.

It was not a good time for that endeavour, nor was Evatt the right person to be making it. The secretary of state, and his policy advisers, along with the armed forces chiefs of staff, were fully absorbed, from March 1946 to June 1950, in building a line of 'containment' in Europe against the further extension westward of Soviet power. The North Atlantic Treaty Organization (NATO) Treaty, which embodied that effort, did not come into force until 1949. China by then had almost fallen to Mao Tse-tung's soldiers, but the south-west Pacific was about the last place on earth in which the United States had any reason to expect a strategic challenge. Besides, Evatt himself was rather 'persona non grata' in Washington. As early as 1944, the Venona decrypts[3] had implicated 'sources close to the Minister' in the leak of information from Canberra to Moscow, and thence to Tokyo.

The advent of the government of Robert Menzies at the end of 1949 removed that particular obstacle, but it was only the outbreak of the Korean War in June 1950 that transformed Washington's strategic stance in the region. In effect, the sudden urgent US need for a peace treaty and continuing security relationship with Japan temporarily endowed Australian policymakers with enough diplomatic clout to enable Percy Spender, the new External Affairs minister, to negotiate the ANZUS Treaty in 1951.

Washington needed Canberra to agree to peace terms with Japan that were considerably less onerous than those originally desired by Australia. At the time, and for most of the following decade, the possibility of a revival of Japanese militarism was a more vivid preoccupation in many Australian minds than the implications of Mao Tse-tung's victory in China. By the early 1960s, however, Menzies found it politically useful to foster an alarming vision of 'Asian communism' as a dagger pointed directly at Australia's heart. After the treaty was signed, Spender became, of his own choice, ambassador in Washington, where he spent much time trying to persuade the Americans to turn ANZUS into something more like a Pacific version of NATO. He was, of course, not successful in this endeavour, with the Americans judging quite rightly that China's neighbours, then as now, were not yet ready to be recruited into the sort of containment strategy that had worked in Europe. Moreover, they also knew, since they were still reading Moscow's diplomatic traffic, that it was Joseph Stalin, rather than Mao Tse-tung, who had encouraged North Korea's bid to take over the south.

The United States therefore interpreted the outbreak of war in Korea as Stalin becoming more adventurous in his world strategy after having acquired atomic weapons in 1949. That was one reason why strategic

---

3   Venona was the codename for a remarkable long-term US success in breaking the Soviet diplomatic cipher system. It began as early as 1943, but was not officially acknowledged to exist until 1995. See Desmond Ball & David Horner, *Breaking the Codes: The KGB's Network in Australia, 1944–1950* (Sydney: Allen & Unwin, 1998), for a full account, with special relevance to Australia.

calculations at the highest official level in Europe and Washington were so extremely pessimistic — not to say alarmist — in the early 1950s. Menzies himself was once reported to have believed that the outbreak of a new world war might come as soon as 1954. Eminent strategists, like General Sir John Hackett, the commander of the British Army of the Rhine, were saying (at least privately) that such a war might mean hundreds of millions dead, mostly in Europe, in the first few days. Many analysts held that atomic or nuclear weapons might spread to as many as 40 countries in a decade or two.

That level of strategic anxiety formed the background to the beginnings of the kind of intellectual evolution of which the history of SDSC is a notable part. I had the good fortune to know some of the founding fathers of this enterprise in London and Canberra quite well — and a few of those in the United States. In my judgement, almost all of them were moved by three common convictions: that war had acquired a new dimension of lethality; that it was far too serious a matter to leave to the generals; and that it deserved a wider level of intellectual enquiry.

Probably the most important influence on academics in Australia at this time was that of their colleagues overseas — especially the influence of those in London concerned with what eventually became the International Institute for Strategic Studies (IISS). I was in London at the time, enrolled in the graduate school at the London School of Economics (LSE), but earning a crust at Chatham House, as rapporteur of a book on Britain and the United Nations (UN). Denis Healey, Michael Howard and Alastair Buchan (three of the most influential 'founding fathers' of IISS) were members of the study group for the book, so I got to know them quite well. All three, as young officers, had been through very hard wars. Healey had been 'beach master' at Anzio (one of the bloodiest landings under German fire of the entire war), Howard had fought through Germany with the Coldstreams, and Buchan, who had been in the Arnhem disaster, once told me that he had seen most of his friends die around him. It was therefore not surprising that all three believed in intellectual effort towards more rational strategies.

The project for an institute more specifically oriented to strategic issues than Chatham House and its overseas affiliates was already being considered — somewhat to the alarm of those in charge at Chatham

House. Yet, in its earliest days as the Institute for Strategic Studies (ISS), it was a small operation, with a director and his secretary, and a couple of rooms in the Adelphi Terrace (where J.M. Barrie used to live). We used to meet over austere luncheons (a sandwich and a glass of wine), yet those were some of the best and frankest seminars I have ever attended. The Whitehall mandarins used to drift up from the Foreign Office, some journalists would drift down from Fleet Street, some colleagues and myself from LSE, and Michael Cook from Australia House. Bill Fox, head of the Institute of War and Peace Studies at Columbia University, was also there at the time, and once told me that the patronage of President Dwight Eisenhower was vital to the rise of such studies at Columbia and elsewhere in the United States. Occasionally, Thomas Schelling, who was to become perhaps the most influential theorist for part of the general enquiry into crises and conflicts, was also in attendance.

Hedley Bull came down from Oxford to teach at LSE just after this period. Though his main interest was always in international theory, his first book, *The Control of the Arms Race* (1961), was in the strategic area, and impressed official opinion so much that he was invited by Harold Wilson (who had just become Labour prime minister, after many years of Conservative dominance) to turn his attention to this subject at the Foreign Office. I was a member of the Advisory Board on Arms Control established by Bull, and attended some of its meetings. Very fascinating they were too, despite the sombre subject matter. Bull had a wonderful knack of getting people from a range of fields to see the relevance of their particular expertise to the possibilities of arms control. So zoologists and seismologists and physicists, as well as international relations people and politicians and diplomats, attended the meetings. I well remember a passionate argument at one of them with the eminent zoologist Lord Walter Rothschild about whether any lessons could be learned from animals on the defence of territory, and other arguments with Patrick Blackett, the equally eminent but very left-wing physicist, about the interpretation of Soviet policy, which most of us thought was pretty threatening at the time. Yet Moscow still had many friends (and some covert allies) in Labour and far-left opinion in Britain. Wilson, and later James Callaghan, as the chief decision-makers for the two Labour governments of the 1960s, had to try to reconcile left-Labour opinion to a policy that was actually of close collaboration with NATO and the United States. One of the ways they accomplished this was by stressing the new initiatives on arms control being evolved at the Foreign Office, with Bull's help. That strategy was later adopted in Canberra, I think, when the then foreign minister (later Governor-General) Bill Hayden faced a similar problem with Labor-left members of the parliamentary party unhappy with the ANZUS connection, just after the Vietnam War. So there is no doubt that Bull had an impact on the formation of official policy, at least while the Labour Party in Britain was in office, and perhaps also in Washington a bit later when Henry Kissinger was in office as secretary of state and the Strategic Arms Reduction Treaty (START) talks were under consideration. I think Bull knew Kissinger quite well — probably from Bull's days at the Foreign Office.

Originally most of the scenarios we contemplated at ISS concerned Europe, but from the mid-1950s we also looked at Middle Eastern problems, especially after the crises in Iran in 1953, and the Suez crisis in 1956. By the early 1960s, we were also becoming preoccupied with those in South-East Asia. In 1967, I wrote for Buchan what I think was the first of the IISS Adelphi papers on the balance (or potential balance) of power in Asia, concluding that its time had not yet arrived. Only very recently, with the rise and rise of China, has the time of a possible future Asian balancer seemed near.

The 1963 IISS conference, held at Pasadena with the help of the RAND Corporation, was, however, almost entirely devoted to the rising crisis over Vietnam, and the question of whether the United States should embark on open combat intervention, as against the covert assistance already being provided to the South Vietnamese. Conferences were very small in those days, with only 29 attending that conference (including five Australians). Buchan always said that we were destined to assume a leading role in international issues in this part of the world. We had various eminent Washington insiders from the administration of President John F. Kennedy talk to us, and they all maintained that the US president, though under heavy pressure, remained determined to avoid sending in combat troops. (That was in August, only three months before Kennedy's assassination in Dallas on 22 November 1963.) The new president, Lyndon B. Johnson, did, of course, yield to that pressure about a year later, after the alleged Gulf of Tonkin incident. I have always wondered whether, if not for Lee Harvey Oswald, the United States and Australia might perhaps have avoided that disastrous battlefield.

Such a possibility would not have been welcome in Canberra at the time. Opinion close to the decision-making level had, for a couple of years, been far more hawkish on Vietnam than official opinion in Washington. Many people took seriously the notion of a 'Beijing–Jakarta Axis'; that is, a sort of revolutionary alliance between the PKI (Communist Party) in Indonesia and the Maoists in China to evict what were called the 'old established forces' (i.e. the Western powers) from this part of the world in favour of 'the new emerging forces' (i.e. the rising tide of left-nationalist insurgency). Such an axis was easy to believe at the time, with Indonesian President Sukarno practising 'Konfrontasi' against the British, Ho Chi Minh looking ever more likely to win in Vietnam, and Maoism at the peak of its ideological attraction. But by 1965, of course, Sukarno was out of power and the United States was heavily involved in South-East Asia for the next decade. Moreover, by 1969, Moscow and Beijing were quarrelling, Kissinger was starting work on his détente strategy, and the balance of forces in the world was beginning to change.

As far as atomic or nuclear weapons were concerned, the late 1960s is the period when the notion of Australia turning to them for defensive purposes lost the appeal it once had for a few policymakers and others. Australians had been involved in the research at Cambridge and elsewhere (long before Hiroshima) that led in time to the Manhattan Project, through various scientists, especially Mark Oliphant. A school of nuclear research had been established at The Australian National University (ANU), and Australia had plenty of uranium. So for some years the project seemed quite feasible. However, after the time of John Gorton as prime minister in the late 1960s (he was a fervent Australian nationalist and rather distrustful of our 'great and powerful friends'), strategic opinion turned firmly against it, despite some high-level lobbying by a few influential insiders. The idea was certainly dead by 1970, as much for cost considerations as from a preference for reliance on the US nuclear umbrella. I believe some academic theorists (both in Australia and at IISS) had a good deal to do with that change of opinion — not perhaps directly, but because they created a climate of opinion that gradually diffused into policymaking and made much clearer (than originally) the dangers and disadvantages of 'going the nuclear route'.

Since 1966, SDSC has been a regular forum for non-official strategic debate in Australia, but I might venture a few words about some scholars in the field who were somewhat outside the mainstream, for one reason or another. Arthur Burns, who held a chair of political science at ANU, was an early entrant into this sphere of enquiry. He published an influential monograph, *From Balance to Deterrence: A Theoretical Analysis*, in 1956, and a more substantial work, *Of Powers and Their Politics: A Critique of Theoretical Approaches*, in 1968. He was a lively and original thinker, who had some input into the foundation of SDSC, and was more at home with mathematical models than most Australians in the field. Some of the insights on the control of conflict have come from scholars associated with peace research rather than with directly strategic issues — notably professors Andrew Mack and Ramesh Thakur on the now influential concept of 'human security' as against 'national security'. A small countercurrent to mainstream strategic analysis, sceptical or distrustful of conventional official assumptions, has tended to suggest alternative modes of meeting security dilemmas, or has even denied their existence. Their influence has waxed and waned in accordance with international events. They were rather influential after the final collapse in Vietnam, and might be again if things deteriorate after the Western pull-out in Iraq (if, for instance, the situation there resembles endemic civil war rather than a budding democracy).

One prospective strategic dilemma (which has perhaps loomed larger in public opinion than among policymakers or their advisers) is whether Australia's status as a close ally of the United States seriously damages its standing among regional neighbours, who are usually disapproving of US policies — especially in the non-West. Does, in effect, current US unpopularity with world opinion rub off on Australia as a regional power, which is conscious these days of its present and future need to get on satisfactorily with the Asian governments — governments that, in turn, are increasingly anxious to promote and define a regional identity?

The answer must I think be formulated on two levels. Firstly, at the level of top official decision-makers and their advisers in our region, the climate of intellectual opinion seems tough-minded and realistic. They are fully conscious of the rise and rise of Chinese power, and know that they must live with it and make their own policies accordingly. Yet they are also conscious that it may be useful to have a global balance of power that will act as a restraining factor on future ambitions that might prove damaging to their own interests. They are therefore not necessarily anxious to see US power opt out of East Asia, and they are fully conscious that Australian security ties are one of the factors keeping Washington interested in this part of the world. So, on the whole, at that level, the alliance may be an asset rather than a hindrance, even though political leaders (like Malaysia's Mahathir Mohamad) have found it a politically useful stick to beat Australia with.

A more complex issue is often present in these calculations: the status of Taiwan. Canberra stands firmly with the United States on its long-established 'one China' policy; that is, the status quo needs to be maintained by both sides. (No unilateral declaration of independence by Taiwan: no military attack from the mainland.) But if the efforts to prevent matters coming to a crisis fail, and the United States should become involved in hostilities over Taiwan, I have heard Canberra's attitude defined to the United States in somewhat cryptic terms as 'we will go up the hill with you, but we will not jump over the precipice with you'. No doubt the circumstances of the time would determine the issue, but a carefully ambiguous signal is often useful in diplomacy. On the whole, the kind of public apprehension often raised in Australia about the US alliance damaging our relations with regional powers seems to me unjustified, especially with regard to China, whose own diplomacy worldwide is impressively sophisticated.

Finally, it might be asked whether the factors that have shaped Australia's strategic culture in the past several decades will be equally influential in the next few. One established factor will, to my mind, certainly persist. Australians are not a militaristic people, but they seem to take an ever-increasing pride in their military traditions. The Gallipoli campaign, as commemorated on Anzac Day, appears to have become the defining icon of our national identity. Federation certainly cannot compete with it, nor can the events of the Eureka Stockade of 3 December 1854, which was a brief and local affair. Historically speaking, Australia is a bit short on drama: no Declaration of Independence and no civil war. For a time after the Vietnam War, the armed forces were unpopular, but that does not seem to me likely to be duplicated after Iraq. We have apparently perfected a technique of sending only very small, highly professional forces that, thankfully, do not incur many casualties but do Australia credit in 'niche roles' with much larger allied forces. I think, therefore, that, on the whole, the tradition of the expeditionary force will persist, and the US alliance will continue to be regarded as a useful insurance policy, which enhances our security and our diplomatic clout at a rather low cost.

In retrospect, strategic debate in Australia can be seen to have focused closely on security preoccupations and sometimes allayed them. For the first five years of the nuclear age, the 'shock and awe' factor kept preoccupations chiefly on nuclear risk. The first task I was given in the Department of External Affairs was to write an analysis of the implications for Australia of the Baruch Plan, the first US effort at the UN to put the nuclear genie (which had been released in 1945) back in its bottle. I reported that there were no implications for Australia, because the Russians were going to kill the plan stone dead in the United Nations Security Council (UNSC), being determined to get nuclear weapons of their own. Unfortunately that turned out to be true. Ever since, my own work has focused on the relationship between strategy and diplomacy, and the way each is the context for the other. The Korean War (1950–53) had the strategic result of (as was said at the time) 'putting teeth into NATO'; in effect turning it from a diplomatic alliance into a forward-deployed military coalition. Surprising at first sight, but actually logical. Even before the assumption was confirmed in the last decade or so by the release in Moscow of the old Soviet archives, it was easy to see the diplomatic advantage to the Soviet Union of averting any rapprochement between the United States and China. In effect, it delayed détente for 20 years, until the Kissinger visit of 1971, and those were years when Moscow did pretty well in the Third World, whilst NATO built up its strength in Europe.

For Australia, however, late 1951 saw the development of what was proved (in retrospect) to be the 'foundation debate' of our foreign policy, whose outcome has governed our diplomatic choices for the whole 55 years since. It was the moment of choice of the US alliance as our 'guiding star' in the larger issues of external relations — those that bear on global conflicts, as against regional relationships with the small sovereignties of our immediate neighbourhood. The balance between those two commitments is the subject of much argument at the moment, and those arguments are likely to continue. No prime minister since Menzies' time has deviated from attachment to the US alliance, though Gough Whitlam in 1973–74 seemed for a while as if he might, and that appears to have been the only moment when the alliance was in some doubt in Washington, though if Mark Latham had made it to prime minister there might possibly have been another.

To revert for a moment to 1951, Spender ought, for good or ill, to be seen as the 'founding father' of the alliance instead of Menzies, who remained sceptical about the whole idea until it was up and running. At the beginning, he was reported to have said: 'Percy is trying to build a castle on a foundation of jelly.' However, the most consistent reaction against the ANZUS Treaty came from those still intensely distrustful of Japan, and fearful that Washington's need for an ally in north-east Asia would lead to a rapid rebuilding of Japan's industrial strength and a resurgence of militarism there. It took a decade or more for the memory of Australian prisoners-of-war emerging from Japanese camps as walking skeletons to fade from Australian minds.

Involvement in the war in Korea did not create anything like the degree of resentment in Australia that Vietnam did 15 years later, perhaps because it was fought under a UN banner, there was a more direct conventional invasion, and it was much shorter. The most notable pre-Vietnam occasion of anti-US fervour came in the early 1960s, over what was then called Dutch New Guinea (later Irian Jaya, and now Papua). When the Kennedy administration came to office in Washington, they swiftly made it clear that they were not going to back Australia's efforts (which dated from Evatt's time) to keep the disputed territory free of Indonesian sovereignty. Without that diplomatic backing, neither Dutch nor Australian efforts had any hope of prevailing, so there was a disgracefully bogus 'Act of Choice' under UN auspices, and the stage was set for the more recent embarrassments. Yet the chances of effective resistance then were, to my mind, non-existent: the Dutch could not fight there, nor ourselves, and the Kennedy administration (in between the mounting problems in Vietnam, the looming crisis in Cuba, and the general Cold War tensions with the Soviet Union) could not really be expected to spare much thought for an obscure corner of the South Pacific.

In retrospect, I think we drew the wrong lesson from the Cuban missile crisis. What it really demonstrated was that, faced with a potential nuclear brink, the great powers will draw back and pursue their hostilities another way. For the United States and the Soviet Union, that meant, for the most part, a continued Cold War; but for lesser powers, or non-state actors, it seems probable that now it will tend to be asymmetric war governed by, more often than not, urban guerilla tactics. However, those concepts have only begun to be developed since 2001, and US Secretary of Defense Donald Rumsfeld apparently did not take much note of them when he sent US troops into Iraq.

The early 1960s were, however, in other ways almost as anxious a period in policymaking circles in Canberra as early 1942. Conditions were clearly deteriorating in Vietnam for its government in Saigon, and there was widespread uncertainty as to whether the United States would venture into combat involvement there, and even idle talk about the use of nuclear weapons. At the same time, the situation in Indonesia seemed to be advancing towards a coup, either by the generals or the communists. President Sukarno had embarked on his Konfrontasi with Malaysia, and consequently was waging a sporadic guerilla war against the British, who were asking for military and diplomatic assistance from Australia. (Our diplomats in Jakarta managed that crisis adroitly.) China, in Mao Tse-tung's disastrous final years in power, was evincing a spasm of revolutionary fervour, and cultivating the Indonesian communist party with particular zeal.

This period can therefore be seen as the first intimation of what I think is going to remain the characteristic dilemma of Australian policymakers in the fields of foreign policy and defence: how to hold the balance between Australia's regional commitments and anxieties and its global ones. Until the end of the colonial world, after World War II, international politics for three centuries revolved around Europe, with Japan and its Asian ambitions making a late entry in the 1930s, whilst the rest of the Third World had hardly any impact. Australia's destiny was thus essentially shaped by global factors, like the two World Wars and the Great Depression. That time (which Asian historians have called the Vasco da Gama era) has passed, but not our involvement with the central balance powers and their relationships. So our strategic debates must take account of both sides of the balance, and sometimes that will make for difficult choices, for instance, as to the type of weaponry chosen and training undertaken.

The world will probably be more turbulent over the next 40 years than in the past 40. That is because a multipolar society of states (which will probably replace the present unipolar one within that timeframe) offers far more opportunities for international friction and crisis than either a bipolar one (as in the Cold War) or a unipolar one (as from 1992 to date).

Australia will be living with a 'company of giants'. Four mega-powers — the United States, the European Union, China and India — will dominate the centre stage of world politics, but there will be at least six other formidable powers whose interests and capabilities will have to be taken into consideration by strategists: Russia, Japan, Pakistan, Mexico, Brazil and Nigeria. World population will have risen towards 9 billion, as against 6 billion at present. China and India together will account for about 3 billion, the other Asians for another billion.[4] So Australia will have about 4 billion neighbours — most of them still quite poor. The United States will run to about 400 million citizens, and still have the most formidable advanced military. The European Union will run to about 600 million, but its power will still be mostly economic rather than military. Among the increasing world numbers, there will be about 2 billion Muslims, a tiny fraction of whom may still be minded to jihad. At least eight governments, possibly 10, will have nuclear weapons. In the coming decades of 'energy insecurity', small powers in the Persian Gulf, Africa, Latin America or Central Asia will acquire more diplomatic clout through possession of oil-bearing real estate, or offshore resources.

It is not only population numbers that need to be taken into account when assessing prospective pressures on resources and the environment; it is the 'revolution of rising expectations' — the determination of the governments and peoples of the poor world to live more like the rich world. With its present rate of economic growth (sometimes 10 per cent a year), China could relatively soon overtake the United States as the largest economy in the world. India might be its closest competitor a little later, for assorted demographic and political or social reasons. As countries grow more affluent, their demand for resources grows exponentially; so, for example, world demand for oil might quadruple as populations grow by 50 per cent. Moreover, supply of that commodity will become far more expensive as less available sources, like tar sands, have to be developed. The $100 barrel of oil is already on the cards. Unless there is a technological

---

4   United Nations, Population Division: Forecasts to End of Century: ESA/P/WP (Department of Economics and Social Affairs, 2003), p. 187.

revolution in the supply of energy, the old conflict between the 'haves' and the 'have-nots' could be re-created on a vastly larger scale than in the 1930s. For the next decade or so, our security preoccupations will no doubt remain focused on 'non-state actors' and their capacity for terrorist operations. In time, however, the traditional anxieties about the great powers (new or old) and their possible ambitions will be back, and the only feasible remedy for those anxieties will be diplomatic strategies and institutions.

All in all, one has to see the rest of this century as a time when Australia will be dealing with more complex diplomatic and strategic problems than it has faced in the past. It is still likely to feel the need for a 'great and powerful friend'. An unnecessary crisis with China is the only occasion that I can see as having alliance-breaking potential, as far as the United States is concerned, and I think that signal has probably already been conveyed to Washington. On a more hopeful note, however, one could also say that the transition to a multipolar world might expand the range of choice. We might aspire to friends, rather than a single friend, and to a multilateral alliance structure rather than a bilateral one. India will be a new power in Asia, whilst Japan will be an old one revived. Their interests will perhaps run parallel to those of Australia. There is no doubt that we will need resourceful diplomatists and strategists in Canberra for this more complex world.

As World War II approached, the Australian prime minister of the time, Joseph Lyons, spoke plaintively about a 'Pacific Pact'. It was not contemplated in any depth at the time, and has not really been since, if one takes the term to mean a truly comprehensive Asia-Pacific security community. In the more numerous, more powerful, and more interconnected group of sovereignties that will emerge in this region over the next 40 years, however, such a 'pact' might again be under consideration. All the powers will have more to lose than ever before. With luck, SDSC may still be around to help untangle their problems.

2

Strategic Studies in a Changing World

T.B. Millar

This essay was previously published in the 40th anniversary edition. It is reprinted here in its near original format.

At the beginning of his excellent book of memoirs, Dean Acheson quotes the 13th-century Spanish monarch Alphonso X, who said that if he had been present at the creation of the universe, he would have had some sound advice to give the Creator for a rather better ordering of things. Having been present at the creation of the Strategic and Defence Studies Centre (SDSC), I cannot absolve myself of any responsibility, but looking at its achievements over these 25 years (1966–91), I can only wonder at the size and strength of the oak that has sprung from the acorn we then planted.

I joined The Australian National University (ANU) when I was in London in July 1962, and arrived in Canberra by train in October. Canberra railway station in those days was, I believe, the original edifice, built out of weatherboards 50 years earlier and painted government brown. It was an appropriate station for a one-horse country town; it was not particularly appropriate for the national capital, but the national capital was only in the early stages of moving from the one-horse country town it had been to becoming the city we now have. The lake bed had been scraped, and the lake was beginning to fill. Those crude barrack-like structures that were to house so much of the Department of Defence had not yet been built. Major government departments, including Defence, had only recently moved to Canberra, with Defence being housed alongside External Affairs in that sunken battleship known as the Administration Building. Mr R.G. Menzies had been prime minister for the previous 13 years, and looked like going on being prime minister for ever, so divided was the Labor Party. The Liberals had had a narrow squeak at the 1961 election, but God was now back in his heaven and all was right with the world.

Well, perhaps not all. Australia still had a battalion and some aircraft in Malaya, although the Emergency there had been declared over. Sukarno was proclaiming the New Emerging Forces, which were to take over from the forces of neo-colonialism. The ferment in Indo-China, which had forced the end of French rule there, was continuing, despite the arrangements made at the 1954 Geneva Conference. The administration of President John F. Kennedy in the United States had put its foot into the quagmire of Vietnam, believing that American power, efficiency and goodwill, from an impeccably anti-imperial background, would settle the mess left by the departing colonial French, and strike a blow for freedom and against communism in the process. Although the term had not gained current usage, Australian security was based on the concept of 'forward defence'; i.e. that Australia should be defended, largely by powerful friends, as far forward of the mainland and with as little cost to Australia as possible. An idea of the size of the cost was the provision in 1962 at American request of a mere 30 instructors to help train the South Vietnamese Army.

I joined the Department of International Relations, at that time housed in a building on Liversidge St. One day in mid-1963 I had a visit from Arthur Lowndes, a member of the then Australian Broadcasting Commission (ABC) and president of the Australian Institute of Political Science (AIPS). He told me that AIPS held a summer school in Canberra every January, and the 1964 conference was due to discuss Australian defence and foreign policy. He was looking for someone to give a paper on 'Australia's defence needs'. I told him that I was not the one. I had left the regular army as a young captain in 1950 and the Civilian Military Force (CMF) as a major in 1953. I was singularly ill-informed on defence questions. The area of my research at ANU was international institutions, especially the United Nations and the Commonwealth. I suggested he try one of the generals. He said that no serving officer would be allowed to give the paper, and they had tried everyone else they could think of. So, *faute de mieux*, I took it on, and the whole course of my life was changed. SDSC's 25th anniversary conference in 1991 stemmed in a direct line from Lowndes' visit and his powers of persuasion.

Writing the paper was a considerable challenge to me. Despite having served eight years in the regular army, I am not by nature a military man; so I consulted everyone I could find. I remember going to Defence and talking with Gordon Blakers and Sam Landau, who were then (I think) assistant secretaries. I had a list of about 20 questions. The two men were friendly and polite, but were able, or prepared, to answer almost none of my questions, on grounds of military security. Defence, in the government's view, was not a matter for public inquiry or debate; the public should simply accept the Defence provisions that the Australian Government, in its superior wisdom and knowledge, provided. I remember asking for the outline of the pentropic division, into which the Australian Army was in the process of being reorganised. They replied that this was classified information. I pointed out that if I was still in the CMF I would be giving lectures on this subject. 'Yes,' they said, 'and you would be subject to military discipline.'

During those few months that I spent writing that paper, I learned my first lesson about getting information which the establishment, for whatever honourable or dishonourable reason, wants to conceal from the public: you don't go to the top, for they won't tell you. You don't go to the bottom, because they never know more than a tiny bit of the picture, and even that is often distorted. You go to the bright people in the middle, the ones who have had some experience of the system, want to see it improved but cannot do much about it themselves. I spoke to everyone who would speak to me, and gradually put together a paper full of splendid sentiments, pungent comments, and proposals for brilliant initiatives. Then one day, as I was reading over the mellifluous phrases with more than a hint of self-satisfaction, a terrible thought came to me: if you were the minister for defence, would you do the things you so fervently advocate? And I realised that I would not necessarily do them. So, for good or ill, I rewrote the paper and gave it at the conference at the Australia Day weekend in January 1964.[1]

Reading it again years later, it seems to me to have not been a bad paper. Bernard Shaw said that he liked to quote himself as it added spice to his conversation. I do not claim his literary felicity, but I believe that most of what I said at that conference still stands up. Among other things, I said that Australia is primarily and ultimately responsible for its own security; that we must produce security if we are to consume it — we must pay our insurance premiums, our club fees. I recommended that we should not be involved in operations in both Malaya and Vietnam simultaneously, but if we were to be able to make a significant contribution in the region we had no alternative to introducing some form of conscription. I was critical of the pentropic organisation, which was splendid in theory but unwieldy in operation, and of the decision to buy the F–111 aircraft, still at that stage on the drawing boards. I recommended that we have a joint services staff college, and a lot of other things. Several eminent people took me to task for my proposals. Malcolm Fraser, then a rising young Liberal backbencher, could not understand why I was not better informed — I only had to ask, he said. General Wilton, Chief of the General Staff, called me in and told me that I didn't know what I was talking about as regards the pentropic organisation, and he caused an article to be written and published in the *Australian Army Journal*, entitled 'King of the Jungle or Paper Tiger?'[2] A few months later the pentropic organisation was abandoned, so presumably it was not the king of our jungle. Air Chief Marshal Scherger, Chairman of the Chiefs of Staff Committee, called me in and told me that I didn't know what I was talking about regarding the F–111. He quoted to me from *Fortune Magazine*; I quoted to him from *Aviation Week and Space Technology*. It turned out to be a good aircraft, but of course we did not get it until long after the time when our political masters had thought it might be needed.

---

1   The paper was subsequently published as 'Australia's Defence Needs' in John Wilkes (ed.), *Australia's Defence and Foreign Policy* (Melbourne: Angus & Robertson, for the Australian Institute of Political Science, 1964), chpt 3.
2   Directorate of Military Training, 'King of the Jungle or Paper Tiger?', *Australian Army Journal*, No. 179, Apr. 1964, pp. 5–9.

The summer school generated considerable public debate on defence matters, more than at any time since World War II. I like to think that my contribution, together with that of the Minister for External Affairs Sir Garfield Barwick, US Assistant Secretary of State Roger Hilsman, and others, helped to make it a more informed debate than it would otherwise have been.

Shortly after the summer school, I took off for my first visit to Asia as an academic. At Canberra airport that afternoon I purchased an evening paper and discovered that the HMAS *Melbourne* had sunk the HMAS *Voyager*, with considerable loss of life and substantially reducing our naval capacity. While I was in India, as I recall, I received a letter from Peter Ryan, the energetic director of Melbourne University Press (MUP). He suggested I write a book on defence, which I did when I returned, and it was published by MUP the following year.[3] As far as I know, it was the first substantial monograph on the subject written by any Australian.

This time, as I did my research, more doors were open to me. Events in Indonesia, Malaysia and Vietnam brought a heightened public awareness of defence questions, and serving officers and civil servants (especially in the Department of External Affairs, as distinct from Defence) were more ready to talk to me off the record. I remember giving an address in Sydney on the defence of Papua and New Guinea. In preparing for the address, I was concerned at the extent of Indonesian military activity in West New Guinea, and talked to a senior officer in External Affairs about it. He got out the relevant file, and quoted to me the intelligence estimate, which I could only use as background and without attribution, but which was quite a low figure — a thousand or two, as I recall. Flying up to Sydney a few days later to give the talk, I found myself seated next to a senior Defence official. I asked him the same question, but he replied that he was unable to comment. Trying for a reaction, I said I had seen (which I had) an estimate of 30,000 Indonesian troops — a wildly exaggerated figure, of course. He would not respond, but I saw him wrestling with his conscience all the way to Sydney. He then gave me a lift into the

---

3   T.B. Millar, *Australia's Defence* (Carlton: Melbourne University Press, 1965). A second edition was published in 1969.

city in his Commonwealth car and, as we neared our destination, he said to me: 'You know that figure of Indonesians you suggested?' 'Yes', I replied. 'It's too big', he said.

Writing a book about defence was a very different thing from giving a paper at a conference. I had never written a book of any kind, and I had none of the techniques, but it was great fun nevertheless and, following the AIPS venture and the considerable public speaking I was called on to do as a result, I had a sense — probably an exaggerated one — that what I was writing was important, and that I had to get it right because, in some way, the security of the nation was involved. Many friends and old comrades in the Services were only too delighted to have me ride their hobby horses, and I was not always as discriminating an equestrian as I should have been; but the book was well received, and proved a boon to the Service staff colleges, which had never had an Australian textbook to work from. One point on which I was criticised was the cover. The dust jacket of the first edition showed Australia with a series of menacing red arrows above it, pointing down. In fact I had nothing to do with the cover. Ryan was a good editor and became a warm friend, but he did not believe in showing authors the dust jackets of their work, in case they objected — as I certainly would have done. The cover was changed for the second edition four years later.

The book was published in late April 1965. Just before copies were shipped off to bookshops, Menzies announced that Australia would be sending a battalion to Vietnam. It was too late to revise the text but, at Ryan's suggestion, I wrote a postscript that was printed separately and inserted as a slip inside the cover. I do not have a copy of the slip, and do not remember exactly what I wrote, except for the last sentence, which was: 'We are paying the penalty for years of neglect.' A bit dramatic, perhaps, but largely true, as it would have been true of many other occasions in our history. Rudyard Kipling's poem on the infantry soldier Tommy Atkins epitomises the British and Australian attitudes to defence:

> For it's Tommy this, an' Tommy that, an' 'Chuck him out, the brute!'
> But it's 'Saviour of 'is country' when the guns begin to shoot.[4]

---

4   Rudyard Kipling, 'Tommy', *The Penguin Poetry Library* (London: Penguin Books, 1977), p. 161.

Our participation in the Vietnam War stimulated public debate on defence matters. It is now largely forgotten how widespread was the support for the decision to enter the war in the early stages, and how easily the government managed to get the precedent-breaking legislation of conscription for overseas service through parliament and the country. My own feeling is that Menzies' almost entirely erroneous description of the Vietnam War as representing China thrusting down between the Indian and Pacific Oceans, played on the 'yellow peril' syndrome that was latent within the Australian consciousness, and was a big factor in winning public acceptance for the war. I remember commenting adversely on his analysis at a meeting of the Australian Institute of Public Administration, and being told by a senior public servant in the audience that Menzies knew a lot more about the matter than I did. Asked in parliament to comment on something I had written, Menzies referred to me as 'some scribbler in Canberra'. If I ever get around to writing my memoirs, I am thinking of calling them 'Memoirs of a Scribbler in Canberra'.

We all no doubt have our memories of these heady days. I felt it was unwise for Australia to get involved in two wars simultaneously — in Malaysia and Vietnam — but once the decision was taken to go to Vietnam I broadly supported it, on the basis of paying our club fees, and also because I felt the people of South Vietnam were entitled to live out their lives in safety. I remember taking part in the first teach-in on Vietnam in Canberra, in the Childers St hall, at which the author Morris West addressed the packed audience in highly emotional tones. The meeting started at about 7.30 pm and went on until the early hours of the morning. I spoke some time after midnight, when the crowd had not noticeably diminished, such was the public interest. I recall saying that nowhere had a communist government taken over by democratic means. Bruce MacFarlane, who was in the audience, shouted out: 'What about Czechoslovakia?' To which I replied, emulating the Duke of Wellington: 'If you believe that, you can believe anything.' He was not pleased with me. A good many academics became involved in the public debate on defence and security questions, centred on Vietnam — I am thinking of people like Bruce Miller, Arthur Burns, Ian Wilson, Arthur Huck, Max Teichmann, Joe Camilleri, Glen Barclay, Harry Gelber, Hedley Bull, Peter Boyce, Brian Beddie, Jamie Mackie,

Greg Clark, Peter King, and Patrick FitzGerald, among others; and journalists like Peter Hastings, Denis Warner, and Bruce Grant; and I also wrote and broadcasted about it.

One ANU academic with a continuing entrepreneurial interest in defence questions was Arthur Burns, formerly of the Department of International Relations and then of the Department of Political Science in the Research School of Social Sciences. Burns set up a Defence Studies Project at ANU, with the encouragement of Professor Leicester Webb and with some support from the Australian Institute of International Affairs (AIIA). The project organised two conferences, one on nuclear dispersal in Asia and the Indo-Pacific region, and one on Commonwealth responsibilities for security in the Indo-Pacific region, in 1965 and 1966 respectively. The proceedings of both were published in reduced format.[5] Burns had a strong interest in the setting up of the Centre, and its continuing activities.

The Centre came about in the following way. One day, in early 1966, Bruce Miller, Professor of International Relations at ANU, told me of the relatively new idea of creating centres or units within the university, separate from or inside departments. It was a way of attracting outside funds, and of concentrating academic activity on a field of interest. Accordingly, at Miller's suggestion and with his help, I worked up a proposal for a centre to study strategic and defence questions. Sir John Crawford was Director of the Research School of Pacific Studies (RSPacS), and I talked it over with him. He undertook to try to get funds from the Ford Foundation for the project, and did so. Miller, who was very supportive, unfortunately went on study leave at a crucial time, and George Modelski became acting head of the Department of International Relations. I did not feel he was as sympathetic to the project as Miller, and the upshot was that, when it was established, it was given its own organisation separate from International Relations. In those days, heads of departments in the research schools went on study leave for one year in four, and I did not feel that I could take the risk of having the new project in unfriendly hands for such a high proportion of the time, especially its inaugural year. I may have been

---

5   Nina Heathcote (ed.), *Nuclear Dispersal in Asia and the Indo-Pacific Region* (Canberra: ANU, for the Defence Studies Project and the Australian Institute of International Affairs, 1965); and A.L. Burns & Nina Heathcote (eds), *Commonwealth Responsibilities for Security in the Indo-Pacific Region* (Canberra: ANU, for the Defence Studies Project and the Australian Institute of International Affairs, March 1966).

doing Modelski an injustice, but I knew he regarded me as dangerously right wing. The proposal was discussed at a faculty board meeting, but Crawford took the decision to create the Centre, with an advisory committee and me as executive officer (as the headship was then called), although I retained my position in International Relations, at that time being a senior fellow. The object of the Centre was to advance the study of Australian, regional, and global strategic defence issues.

Before getting the Centre underway, I went to Britain and the United States in order to see how they managed these things. In London I talked to Alastair Buchan at the Institute for Strategic Studies (ISS), to Michael Howard, Professor of War Studies at King's College, London, and to a private luncheon at the Royal Institute of International Affairs, Chatham House, to which Foreign Office and Defence Ministry officials were invited. In the United States I spoke with a number of the defence think tanks. None of them provided a clear model for us to copy. The various institutes had access to substantial private funding, which (despite Ford Foundation generosity) we could not feel confident of, on any continuing basis. King's College had a Department of War Studies, as a formal academic activity of the university. We were not ready for that, and we were not into military history, which was a substantial part of the King's College curriculum.

We gave much thought to the question of access to, and use of, classified information. Crawford was strongly opposed to seeking such access. I suppose I had the rather naive view at that time that classified information was likely to be more accurate than unclassified information. This view underwent modification the longer I stayed in Canberra. Formally, we were not entitled to receive anything classified Restricted or above but, in fact, as Des Ball was to find to his benefit, there are a lot of people holding positions of confidence who are prepared to spill the beans in private, without attribution. The problem is of course that, without access to the same or equally good alternative sources, you cannot always tell the quality of the beans. The British and American institutes obviously had formalised as well as informal access to classified information.

In Australia, we had no tradition of academics having access to confidential government information in the social sciences, almost no tradition of academics writing on defence matters, and only very little experience of academics writing on foreign affairs, the AIIA providing the shining exception. Australian academics writing on political matters tended to be left of centre. This did not cause problems during the short periods of Labor government but, apart from the Depression, and the period 1941–49, Australia until 1972 was almost continuously under conservative rule. The Australian political system, irrespective of the party in power, has tended to be obsessed by secrecy, and conservative governments especially so. The combination of these factors made it difficult for anyone in academia to write on defence without arousing that paranoia with which our politics is so generously endowed.

In the formation of SDSC, we took the decision that we would not seek access to classified information, but that we would be prepared to have comments, from friendly officials, without obligation on either side, on what we wrote. At ANU we had developed good relations with people in the Commonwealth Department of External Affairs, largely because of an initiative taken by Miller, with the help of Sir Alan Watt (a former secretary of External Affairs, who was a visiting fellow in the Department of International Relations) in founding in 1963 what was initially called the 'Third Monday Club'. This was a group of senior diplomats and academics with an interest in external matters, who met over dinner in the Scarth Room at University House on the third Monday of each month. One of us would open the batting on a matter of current interest, and there would be general discussion, entirely off the record. At first the officials approached this rather gingerly, but they came to see that none of us had horns, or an obviously direct line to the Kremlin, and some very good discussions resulted. It was good also from the university's point of view, as our group contained people from several disciplines. Later I was to found what we called the 'Foreign Affairs Club', with relatively senior officials from Defence, External Affairs, Prime Minister and Cabinet, Trade and so forth, and a range of academics, who met over lunch at the old Hotel Canberra, with a guest speaker. These meetings had the additional benefit that — to my astonishment — they introduced officials from different departments to each other. I had not appreciated how compartmented the public service was.

By these means, the ignorance, suspicion and distrust that had existed to a varying degree between the public service and academia were significantly (if selectively) broken down. On our side, of course, we had to watch out for a different danger: that we would become the captive of the establishment, sharing their assumptions and accepting their conclusions. A university that does not have a tradition of dissent is not a proper university. Having Crawford as director had many benefits, but he never quite left the public service, psychologically.

It took several years for the Centre and its work to be generally accepted within ANU as a proper academic activity. I soon realised that some of my colleagues, and academics in other universities, regarded me as (to quote the communist jargon of the time) a 'running dog of the imperialists', or of the Central Intelligence Agency (CIA), or an agent of the Defence Department, but in any case rather disreputable. I am grateful that I did not have the kind of disruption that Bob O'Neill was subjected to later on, but people did say and write some very rude things about me, often with a grain of truth in them, to make them stick. Miller, Crawford, and Oskar Spate who succeeded Crawford as director of RSPacS, were all supportive and, indeed, protective, but I think it was not until well after the Vietnam War was over that the Centre was accorded general respectability. We were of course conscious of the problems, and laboured to ensure that all reasonable points of view were presented at our gatherings, not out of self-protection but because we believed that that was what universities were about.

I remember being rung one day by the ABC, which was arranging a television debate on a planned moratorium march. 'We want you to take the position opposed to the march', the gentleman said. I said that, whereas I thought it likely that my views would be significantly different from those of some of the marchers, I believed strongly in the right of people, whatever their beliefs, to engage in such public protests, and that I thought that it was to protect such rights that we had armed forces. 'Well you'd be no good to us,' said the ABC man.

The Australian Labor Party came to realise that the Centre offered a forum for them to express their dissenting views, but interestingly enough they also came to realise that we provided an expertise on which they could draw. Later, when they became the government, they invited me to chair the Committee of Inquiry into the Citizen Military Forces and the Army Cadets.[6]

---

6   See *Committee of Inquiry into the Citizen Military Forces [Chaired by T.B. Millar] Report*, Mar. 1974 (Canberra: Australian Government Publishing Service).

The name of the Centre was a bit of a mouthful. I tried out various simpler combinations, like 'Defence Studies' or 'Security Studies' and O'Neill (when he took over in 1971) wrestled with the same problem. But no other title fitted exactly what we wanted to study, which was a combination of strategic questions and the problems of national defence. So 'Strategic and Defence Studies Centre' it became, and remained.

To start with, all I had was a secretary. Then, with Ford Foundation funds, we were able to appoint a research fellow. This was Ian Bellany, a nuclear physicist who had worked on arms control and disarmament questions with the British Foreign Office. He was to prove an excellent colleague, and wrote a valuable book that Sydney University Press published in 1972, *Australia in the Nuclear Age: National Defence and National Development?*[7] Ian was a very effective acting head of the Centre when I went on study leave in 1968–69. At the end of his time with us, he took a post at Lancaster University, where he held a chair in international relations.

The Centre was small, but active. Our first conference was in September 1967, on the implications for Australia of Britain's decision to withdraw from east of Suez. In 1968, in conjunction with Chip Wood, director of ANU Press, we launched the Canberra Papers on Strategy and Defence. Alex Hunter, an economist, wrote the first paper, on oil in Australia's defence strategy,[8] and Geoffrey Jukes,[9] Jim Richardson,[10] and Ian Bellany[11] followed. These constituted the beginning of the sizable library of literature on strategic and defence matters for which the Centre has been responsible. O'Neill instituted working papers (which provided quick publication of seminar papers) and reference papers, as well as greatly extending the publication of Canberra Papers.

---

7   See Ian Bellany, *Australia in the Nuclear Age: National Defence and National Development* (Sydney University Press, 1972).

8   Alex Hunter, *Oil Supply in Australia's Defence Strategy*, Canberra Papers on Strategy and Defence No. 1 (Canberra: ANU, 1968).

9   Geoffrey Jukes, *The Strategic Situation in the 1980s*, Canberra Papers on Strategy and Defence No. 2 (Canberra: ANU, 1968).

10   J.L. Richardson, *Australia and the Non-Proliferation Treaty*, Canberra Papers on Strategy and Defence No. 3 (Canberra: ANU, 1968).

11   Ian Bellany, *An Australian Nuclear Force*, Canberra Papers on Strategy and Defence No. 4 (Canberra: ANU, 1968).

In 1968 we had our first international venture, at a conference in Wellington on Australian–New Zealand defence cooperation. By this time the formidable Hedley Bull had joined the Department of International Relations, and the Advisory Committee of SDSC. Victoria University in Wellington provided us with the venue for the conference, and New Zealand academics and officials took part. Australia was less well represented at the official level, due to government caution. The proceedings of this conference were also published,[12] and this became the custom.

In 1969 Watt retired as director of AIIA, and I was appointed in his place. By this time I was a professorial fellow in International Relations, as well as being in charge of the Centre. I realised, and I think other people did also, that these three activities were rather more than one person could adequately handle. Around this time O'Neill joined International Relations, from the staff of Duntroon, and I asked him whether he would be interested in taking over the Centre. O'Neill initially said 'No', but Bull and I persuaded him to let his name go forward for it. I duly resigned in 1971 and O'Neill took over. I believed then, as I do now, that this was the best thing I could have done and one of the best things I have ever done. SDSC has never looked back from that time.

O'Neill's first major — and vital — achievement was to get the Centre within the ANU budgeting system. The Ford Foundation funds ran out, and ANU or other funding was necessary if SDSC was to survive. The existence of the Advisory Committee, which we had set up at the beginning under Crawford's watchful eye, was a help. He (as director) was the first chairman, and this principle has continued to be practised, but for Advisory Committee members we have called on other departments within Pacific Studies and other research schools including Social Sciences and Physical Sciences. After a time, defence specialists from outside ANU came to be members of the Advisory Committee. I tried to interest External Affairs and Defence in the idea, but both wanted to remain detached. No representation meant no responsibility. Nevertheless, Sir James Plimsoll, who was Secretary of External Affairs at the time, was especially helpful, while Defence was kindly disposed if cautiously inclined. It later sought to have

---

12   T.B. Millar (ed.), *Australian–New Zealand Defence Co-operation* (Canberra: ANU, 1968).

a representative on the Advisory Committee, but by then the committee opposed the idea. I am sure that the range of interests represented in the Advisory Committee and the high calibre of the members have served SDSC well, especially in having the Centre firmly established as a respectable part of ANU and funded within the ANU budget. Here I should pay tribute to the support given to the Centre in all sorts of ways by the business manager of the joint schools, Peter Grimshaw.

One of the Centre's first activities stemming from a relationship with Defence was an exercise in futurology, which the Army asked us to undertake. We had a series of meetings of an ad hoc group from several disciplines within ANU trying to peer 10 and more years ahead to make rational guesses as to what kind of world Australia — and the Australian Army — might find itself in. Due to the fact that my secretary of the time could not easily read my writing, this group became known as the Forecastry Group. I do not know to what purpose our conclusions were put, but it was an interesting exercise and I think the Army found it worthwhile. (They could at least blame us if they got it wrong!)

O'Neill was head of SDSC from 1971 to 1982 and, during that time, he extended and expanded its activities considerably, making it one of the most significant and respected institutions of its kind. He developed excellent relations with government, while maintaining the Centre's independence as an academic institution. In his first couple of years, SDSC was shaken increasingly by public opposition to the Vietnam War, which found its echoes or indeed its voice in academia. One visiting American lecturer was prevented from speaking, and gangs of student thugs threatened to smash up the SDSC library. Now that I am retired, I feel I can say that some ANU senior officers displayed a regrettable lack of intestinal fortitude in dealing with these disrupting elements, and the normal activities of the Centre suffered. It had been our custom to give briefings to visiting groups from the Royal College of Defence Studies in London, and the US National Defense College. SDSC was required to do this off-campus.

O'Neill sought and obtained funds from the Ford Foundation. The small library resources capacity was extended, with a full-time librarian/ research assistant. A visiting fellows program was initiated, bringing academics and occasionally officials from various parts of Asia and the Pacific for specific research. In 1978 this was supplemented by funds from the Department of Defence for a Defence Visiting Fellowship, occupied either by a serving officer or a civil servant. In 1984, under a separate agreement with the Navy, a naval fellowship was established. In 1985 Defence agreed to fund two visiting fellowships per year for academics to carry out advanced research on strategic and defence problems. Exchange arrangements were also made with the Institute of International Studies in Beijing, and the Institute on Global Conflict and Cooperation of the University of California, in 1983 and 1985 respectively. The formal staff of the Centre were also increased from time to time so that, by 1991, it had six academic posts — a professor as head (Paul Dibb), a special professor (Des Ball), four senior research fellows or research fellows, four visiting fellows with at least a one-year contract, plus six supporting staff. At any one time there were maybe another two short-term visiting fellows. There was also a part-time research assistant and a part-time clerical officer.

Although still comparatively small by some international standards, this is of course the largest centre of its kind in Australia, or indeed in the southern hemisphere, and its output and influence as a national institution stand comparison with any non-governmental centre anywhere in the world. It may seem surprising that this small country of Australia, which over recent years has become marginal to most international events, should have produced so significant a centre. I see this in some ways as parallel to what has happened in the Australian departments of Foreign Affairs and Trade (DFAT) and of Defence, whereby the considerable professionalism built up over the 30 years of international tension following World War II in which Australia was involved are now devoted less to national survival under threat and the reform of international society than to the application of energetic thought to appropriate pressure points — a kind of political acupuncture.

Yet SDSC, as an academic institution, is rather less — and more — than that: less, in that it does not carry governmental weight, and more in that it has an ongoing educational role, demonstrated not only through the annual conferences, the massive list of publications, the participation of SDSC staff in public lectures, broadcasts, giving evidence to parliamentary committees, contributing to government inquiries, lecturing to and advising the various Defence colleges and so forth, but also in the graduate program that has developed since 1984, first by having doctoral students, and then through the Master's and graduate diploma courses (originally funded by the MacArthur Foundation in Chicago). Students in 1991 came from within Australia, from China, Japan, South-East Asia, and the south-west Pacific.

One aspect, which began in my time, was greatly extended under O'Neill's headship, and continued under Ball's, was the development of studies concerning the defence of Australia. These became especially relevant after the end of the Vietnam War, when it was obvious that there had to be a seachange in Australian attitudes to regional security. With the departure from the region of Australia's 'great and powerful friends', old concepts of 'forward defence' were no longer relevant.

I do not want to write the Centre too high, but the fact is that, for those Australians inside or outside the Defence establishment who are interested in defence matters, SDSC supplies an opportunity for acquiring information and for contributing to the public debate that nowhere else provided.

O'Neill left SDSC in 1982 to become the highly successful director of IISS. Much against my better judgement, I took over the headship again for a couple of years.

I was fortunate, and grateful, to be able to hand SDSC over to someone who had already made a major contribution to the Centre's work, and to the international strategic debate, especially in nuclear missile questions. I hope Des Ball will forgive me if I mention my recollection — and recollections are not invariably accurate — of a tall young man looking rather like an early version of the prophet Moses with sharp eyes peering out from a formidable facial foliage, with bare feet, walking the Coombs corridors. This was the recent university medallist in economics beginning his controversial career in strategic studies with a PhD under Bull's tutelage. Ball was born to controversy as the sparks fly upwards, holding the supportable view that the government was obsessed with secrecy and that many of the things it wanted to keep to itself were the proper and indeed necessary subject of public and informed debate. Being from an earlier generation, with a military background, I was occasionally concerned that some of Ball's revelations might in fact be better not revealed, in the national interest, but I had no doubt of the integrity of his research and its findings. SDSC continued to prosper and increase under his guidance and no one can today doubt that he has filled a unique niche in the public debate on defence matters. I was also delighted when Dibb, a very old friend who has made major contributions to the study and analysis of defence matters, both within the establishment and within ANU, was appointed to replace Ball as head of SDSC.

I here mention three people who have played, in their own ways, vital parts in the work and success of SDSC: Colonel Jol Langtry, who looked after the administrative aspects for many years; Billie Dalrymple, who was secretary to O'Neill, to me, and to Ball; and Elza Sullivan, who ably performed so much of the Centre's word processing from which its considerable list of publications have ensued. In his submission to the review of the Centre in April 1987, Ball wrote that SDSC is 'one of the most successful academic enterprises in the University'. This view is widely held at ANU.

I believe that centres concerned with peace research must be included in any consideration of 'strategic studies'. During the period of my second headship of the Centre, the Labor Government under Bob Hawke, and with Bill Hayden as Minister for Foreign Affairs and Trade, expressed an interest in establishing peace studies at ANU. I tried to get them subsumed within the Centre, but I fought a losing battle against the Advisory Committee (which did not give me the support I expected), the Minister for External Affairs (who wanted to be seen to be doing something about peace), and the board of the Institute of Advanced Studies (which bought the notion that a centre devoted to the studying of peace would somehow bring more peace about). (As Kenneth Boulding once said, this was equivalent to saying that a centre devoted to the study of garbage disposal would lead to an increase in the quantity of garbage to be disposed of.) As is now history, the government decided to establish a separate Peace Research Centre (PRC) within RSPacS in 1984 with a continuing grant from DFAT. My one contribution in this area lay in appointing Andrew Mack as senior research fellow in the SDSC, from which he went on to the even greater heights of head of PRC and then to the chair in International Relations.

The PRC was organised very much on the lines of SDSC, though of course with an emphasis on studying conflict resolution and reduction more than conflict itself. In 1991, PRC had an advisory committee also chaired by the Director of the Research School of Pacific and Asian Studies (RSPAS) and with representation from other departments and the SDSC at ANU, other universities, DFAT, and the Law Reform Commission. It had a head, three senior research fellows, two research fellows, five research assistants, and two other staff. Under Mack's energetic direction, PRC became an important member of the family of such centres around the world. It ran seminars and conferences, produced the quarterly journal *Pacific Research*, and published a number of major monographs, and an astonishing number of working papers. It also had a valuable resources centre.

The introduction of peace studies at the University of Sydney grew out of staff members mainly within the Department of Government, especially Peter King, and a staff–student committee. Given initial financial support from the University of Sydney, a centre was launched by the Minister for Defence in May 1988 and named the Centre for Peace and Conflict Studies (CPACS). Its declared aim was to 'promote the study of conflict prevention and resolution and the long term conditions for peace'. CPACS held seminars, lectures and conferences, including a seminar series on 'Deconstructing Deterrence', the papers from which were subsequently published as *Beyond Deterrence*.[13] A three-day workshop on conflict resolution was run in cooperation with the Law Council of Australia. There was a steady flow of international visitors to CPACS. The University of Sydney provided funds for the first 15 months, covering a half-time secretary and maintenance, plus a grant to purchase an American archive.

Deakin University has had an interest in strategic studies since its inception in 1977, stemming in good measure from the interests of Francis West, teaching courses that included them at undergraduate and graduate level within the School of Social Sciences. From 1991, Deakin offered higher degree courses specifically in this area: a Diploma in Defence Studies and a Master's degree in the same subject. The courses were designed to appeal especially (but by no means only) to serving officers of the armed forces, who received credit for any staff college or other professional military qualifications. Deakin is a leader in the field of 'distance education' in Australia, and it is thus especially useful for officers, public servants, or others who want to study in this area without leaving their normal work environment.

When the Australian Defence Force Academy (ADFA) went into the business of graduate study, it was natural that it would establish a program in defence or strategic studies. The Australian Defence

---

13   *Beyond Deterrence: A Multifaceted Study* (Centre for Peace and Conflict Studies, University of Sydney, 1989).

Studies Centre (ADSC) was accordingly set up in 1987 to promote research and study in all aspects of Australian defence, to support and assist postgraduate and honours degree students at ADFA working in the relevant fields, and to cooperate with other organisations involved in the study of Australian defence. It provides for part-time as well as full-time study. The Master of Defence Studies course is open to civilian as well as military personnel. The choice of subjects includes politics, history, geography, civil engineering, computer science and economics. ADSC has seminars and conferences, and produces monographs. In 1991 it ran a major conference on naval power in the Pacific. It has a steady flow of visiting fellows from within or outside Australia. It has close relations with the study centres established within the three Services: the Air Power Studies Centre, the Maritime Strategic Studies Project, and the Directorate of Army Studies; it also provides consultants to work on study projects identified by the Services. In addition to these three centres, the Australian Defence Force Warfare Centre was established in late 1990 and is based on the old Australian Joint Warfare Establishment at Williamtown.

One member of the university strategic studies/peace studies centres was the Indian Ocean Centre for Peace Studies, sponsored by the University of Western Australia (UWA) and Curtin University, and funded jointly by the federal Department of Employment, Education and Training and DFAT. Drawing on both universities, and multi-disciplinary in character, the centre focused on arms control and the geopolitical setting; environmental, resources and developmental issues; and social justice, equity and the law. The centre stemmed from UWA and Murdoch University academics' long interest in Indian Ocean affairs, and Curtin's Centre of Indian Ocean Regional Studies.

Apart from the various departments of politics, government, or at the Department of International Relations at ANU, and individual academics such as Harry Gelber at the University of Tasmania, Fedor Mediansky at the University of New South Wales, and Joe Camilleri at La Trobe University, the only other Australian university in 1991 with a centre whose activities border or touch on strategic, defence or peace issues is Griffith, especially the Centre for the Study of Australia–Asia Relations (CSAAR). Among its other, more politically oriented studies and publications are the reports of international conferences on Indochina and the prospects for conflict resolution there. In 1991

CSAAR held the major conference 'Security in the Asia-Pacific Region: The Challenge of a Changing Environment', which I had the privilege of attending.

Outside the Defence establishment and the universities, the only institution as at 1991 with even a hint of 'think tankery' was the 'private, non-profit-making' Pacific Security Research Institute in Sydney set up in April 1989, with Owen Harries as President and former diplomat David Anderson as Executive Director. Funded in roughly equal proportions by a number of leading Australian companies and three American foundations, its purpose was to undertake research and stimulate public discussion on foreign, defence and economic policy issues in the Asia-Pacific region with particular reference to Australian national interests. It published several papers in a series called Australia and Tomorrow's Pacific, a paper on Australia's response to the Gulf War (1990–91), and papers delivered at two conferences.

Separate again from any of these institutions and from government is the Royal United Services Institute of Australia, catering largely to serving and retired officers, who have used their state branches and publication of a journal for several decades to keep up a professional debate on the major strategic and defence issues of the day. The Naval Institute is a similar body, with an accent on maritime matters.

Having been out of Australia for much of the six years prior to 1991, I hope that I have not overlooked the work of any institution or individual during that period.

Looking back between 1966 and 1991, it is hard to imagine we are in the same capital, or even the same country. I remember a seminar we ran on the future of the aircraft carrier, with a paper from a student. Around the table in Seminar Room B in the H.C. Coombs Building were something like a dozen admirals and senior captains, none of whom was prepared to say a word on the subject or even to ask a question. Well, at least they were there. Today we would find them actively participating in the discussion, with perhaps one of them giving the paper. The number of books and journal articles on strategic and defence questions has grown exponentially. Strategy and defence are now respectable subjects of academic courses and research. The media have become much better informed — certainly they have a much greater range of expertise to call on.

It is impossible to know what impact this growth of public information and discussion has had on specific policies of government. A number of the things I recommended in that 1964 AIPS paper were implemented not, I believe, because I recommended them, but rather because they were sensible things to do, although sometimes governments will only do sensible things if the public urges them. At times, SDSC seemed to have a more obvious influence. I am thinking here particularly of the 1976 conference on the future of tactical air power in Australia, which was reported at the time to have influenced the government over the principles of selection for a replacement of the Mirage fighters, and also of the Centre's contribution to the Regular Officers Development Committee at around the same time. Certainly, the report of the CMF Inquiry committee,[14] and Paul Dibb's review of Australia's defence capabilities[15] — although neither was an SDSC activity as such — were important parts of the decision-making process. Between 1966 and 1991, the quality and the quantity of debate, and academic contributions to it, changed beyond recognition.

It is tempting to believe that this splendid development has come at a time when it is less needed than ever, with the end of the Cold War, and the reduction of tension within our region. I think that is a profound mistake. Although there is no enemy at our gate, many parts of the world are in a state of tension or even conflict, some of which could spill over into areas or situations affecting Australia. I am not in the business today of drawing up scenarios for the possible deployment of the Australian Defence Forces, as I did in 1964. But perhaps if there had been more people then with a serious professional interest in these matters, Australia might have gone in different directions. And, as I have said many times over these intervening years, threats to our nation and people and way of life can develop more rapidly than we can build up the forces and the philosophies and the expertise to cope with them. At a time when the forces, for economic reasons, are being inevitably decreased, it is heartening to find that the philosophies and the expertise of strategy and defence studies are being given far more and far better attention than at any period in our peacetime history.

14   T.B. Millar, *Committee of Inquiry into the Citizen Military Forces* (Canberra: Australian Government Publishing Service, 1974).
15   Paul Dibb, *Review of Australia's Defence Capabilities*, report to the Minister for Defence (Canberra: Australian Government Publishing Service, 1986).

Plate 1 Dr T.B. Millar, head of SDSC, 1966–71, 1982–84

Plate 2 Professor Sir John Crawford and Professor Anthony Low, chairmen
of the SDSC Advisory Committee, 1966–67 and 1973–75 respectively

Plate 3 Professor J.D.B. Miller, head of the Department of International Relations, 1962–87, and member of the SDSC Advisory Committee, 1966–87

Plate 4 Professor Hedley Bull, joint head of the Department of International Relations and member of the SDSC Advisory Committee, 1967–77

Plate 5 Robert O'Neill, head of SDSC, 1971–82

Plate 6 Colonel J.O. Langtry, executive officer, and Desmond Ball, research fellow, 1975

Plate 7 Joli and Peter Hastings, former research assistant and senior research fellow, May 1986

Plate 8 Robert O'Neill (right) with Gunther Patz, former PhD student, 1992

Plate 9 Colonel J.O. Langtry, executive officer, October 1983

Plate 10 Billie Dalrymple, secretary; Desmond Ball, deputy head; and Ram Subramanian, visiting fellow, Thredbo, 1982

Plate 11 Ray Funnell, Robert O'Neill, Suzanne Funnell,
Billie Dalrymple and Mara Moustafine, 1988

Plate 12 Paul Dibb, senior research fellow, 1984

Plate 13 Desmond Ball, head of SDSC, at the entrance
to the Pine Gap station, July 1984

Plate 14 US President Jimmy Carter and Desmond Ball, April 1985

Plate 15 Desmond Ball and Australian Federation of University Women, Queensland fellow Samina Yasmeen at SDSC, 1985–86

Plate 16 Benjamin Lambeth and Ross Babbage, senior research fellows, and Ray Funnell, July 1986

Plate 17 Mr R.H. Mathams, member of the
SDSC Advisory Committee, 1985–89

Plate 18 Andy Mack, senior research fellow, 1983–85

Plate 19 Greg Fry, research fellow,
1983–86

Plate 20 Dr Lee Ngok,
senior research fellow, 1985–88

Plate 21 Andrew Butfoy, SDSC's
first PhD student, 1984–88

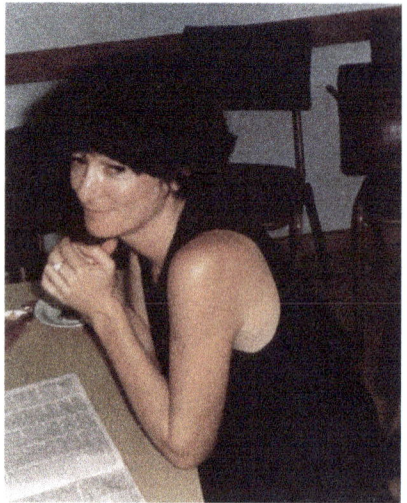

Plate 22 Nicola Baker, PhD student,
1995–2000

3

Strategic Studies in Australia

J.D.B. Miller

This essay was previously published in the 40th anniversary edition. It is reprinted here in its near original format.

Having had the pleasure of reading the chapter by Tom Millar, I endorse every word of it, and congratulate him upon it. Perhaps I can add a few footnotes before I say something of our original hopes for the Strategic and Defence Studies Centre (SDSC) and the climate of opinion within which it was established.

First I want to claim just a little credit for myself. I do not think Millar knows that it was I who told Australian Institute of Political Science (AIPS) President Arthur Lowndes to ask Millar to provide that groundbreaking paper at the 1964 summer school. In that paper, Millar says he was 'singularly ill-informed on defence questions', but there was no doubt in my mind that he was better informed than any other academic in Australia at the time; we had had various informal conversations about military matters, and I was greatly impressed by the acuteness of what he had to say.

The paper itself was a triumph. When it was over, and the cheering and clapping in the Albert Hall were at their height, I remember turning to the person next to me, and saying gleefully, 'It worked!' He or she did not know what had worked, but nodded kindly and said, 'Yes, it must have'.

That paper by Millar led, in a way, to the establishment of the Centre two years later. As Director of the Research School of Pacific Studies (RSPacS), Sir John Crawford was delighted that we had a staff member who could not only talk sensibly and constructively about defence, but who could also excite public interest and maintain it at a high level. When Crawford and I began to talk about the possibility of a Centre, it was a plus point that we already had someone around whom it could be built.

Now let me say something about the climate of opinion in which the idea of the Centre originated. If I may put it in personal terms, when I came back to Australia towards the end of 1962 after 10 years in England, I was appalled at the lack of contact between academics and officials in such fields as defence and foreign affairs, and by the surly and often bloody-minded approach of the governing politicians towards the opinions of those academics who did express themselves in public. It was true, as Millar has pointed out, that much of such comment was from a left-wing standpoint; but that did not excuse the malignity of the political reactions. In England I had become accustomed to public affairs being discussed vehemently but courteously; I had taken part in radio discussions at Chatham House in which people had hit hard but had preserved their tolerance; I had been an active participant in the debate over Suez and had got some hard knocks in exchange, but had lost no contacts as a result. What I found here was a situation in which some academics and the government were making hysterical noises about that paper tiger, the South-East Asia Treaty Organization (SEATO), as if it mattered; in which the university had been convulsed over whether it should award an honorary degree to the King of Thailand; and in which historian C.P. Fitzgerald was still being persecuted, as he had been when I left the country in 1952, over his belief that Australia should recognise communist China.

There seemed to me to be a need for two particular areas of study. On the one hand, there was the strategic situation that affected Australia (one which, contrary to the opinions of many others in Canberra, I believed encompassed the global strategic confrontation between the United States and the Soviet Union, and not just the state of things in South-East Asia). On the other hand, there was the problem of communism as a worldwide phenomenon: how different was it from country to country, in what ways could it be regarded as a threat to Australia, how did it affect our alliances and connections, and what we should do about it. I hoped we could build something around Harry Rigby to extend this kind of study but, by the time I got back from the study leave that Millar mentions in his chapter, Rigby had been, legitimately, snapped up by the Political Science Department in the Research School of Social Sciences, where he pursued a distinguished career.

I do not want to claim too much prescience for these two notions; they both grew out of the situation at the time. Australia had worked itself into a state of mind in which any presence of communism could precipitate a military intervention as in Korea and Malaysia, and later against Indonesia and in Vietnam. The military establishment, having operated continuously abroad since 1939, was committed to 'forward defence', both out of custom and because of any knee-jerk response by politicians to whatever they were told by British and American intelligence was an imminent communist threat. The problems were compounded by the often childish and bitter reactions of the left wing in Australia to any American policy in Asia.

As Millar has mentioned in his chapter, a means that I hoped would relieve tensions between academics and officials was the 'Third Monday Club', which, if memory serves me, we began some time in 1963. I had spent a semester at Columbia University in New York before coming back to Australia, and had been much impressed by a program called the Columbia Seminars, which consisted of monthly dinner meetings at which academics, officials, journalists and businessmen met to consider some subject of public importance. Debate was free, confidentiality was preserved, the standard of membership was high, and the whole operation seemed to foster better understanding between the kinds of people involved.

If I were to reproduce the Columbia Seminars in Canberra in respect of external policy, it was necessary to have firm bases on both the academic and official sides. Here I was fortunate in having the continuing support of Crawford on the one hand and External Affairs Secretary Sir Arthur Tange on the other. Crawford was an academic before he was a public servant, and never forgot it, while Tange was an academic manqué. After discussions with both of them, it became clear that having journalists at the meetings would be too risky, given the general governmental suspicion of them, and that businessmen were not to be found in Canberra — not of any consequence, at all events.

What we finished with was an equal number of academics from The Australian National University (ANU) and of officials (drawn from External Affairs (as it was then), Treasury, Trade, Defence, and Immigration). The officials were from the top levels. We were not going to talk solely about defence, but also about economic and political approaches to the outside world.

The Third Monday Club kept going for 10 years, which meant that it persisted throughout the Vietnam War, a traumatic period during which it might have disintegrated. It is a tribute to the changing membership that it kept going, and that none of its full and frank discussions were ever reported in the newspapers. I am sure it did much to reconcile officials to the idea of academic integrity.

I have talked about the Third Monday Club at some length because I think it was relevant to the success of the Centre when that was established. By that time the ANU and governmental people involved had got to know each other (sometimes on very friendly terms) and were at ease in each other's company, which was a help when we came to set up something that Australia had not seen before — an academic body that would try to be objective about the uses to which our Defence money was put, and the kind of world in which those uses might, or might not, be applied.

Why was it important to establish such a Centre? I think Millar has covered that effectively in his chapter. Defence was still a sanctum sanctorum. The forces had been kept under effective civilian control for a long time, but it had been, in effect, the result of an implied agreement between civilians and the military that the public need not be informed of what was going on, and that the air of public debate should be filled instead with windy rhetoric about China. We did not know then it was windy rhetoric; there was indeed some evidence that, by implication, suggested there might be something in it. But the upshot was that defence itself was discussed only perfunctorily in parliament, and that the major issue of nuclear warfare was hardly discussed at all. There were people in Australia — nuclear physicist Ernest Titterton and chemical engineer Philip Baxter foremost amongst them — who wanted us to produce nuclear weapons; the issue was rarely discussed in public, and there was an urgent need that it should be. The question of forward defence was so bound up with that of the American alliance that a rational examination of it was hardly possible. In all, there was a great need for an institution that would give these questions a proper scrutiny.

Now let me say how much SDSC owed to Crawford. Crawford was a complex man (as I tried to show in the book that Lloyd Evans and I edited about him)[1] who (as Millar noted at the SDSC's 25th anniversary conference) 'never quite left the public service, psychologically'. But he did believe in open discussion, and he did have a personal interest in defence policy. In ways that he hinted at, but which I never pinned down, he had been associated with defence and intelligence questions when he was a permanent head. I think, though I cannot swear to this, that it had made him sceptical of intelligence assessments, and he was convinced that in this field, as in assessments of economic policy, there should be independent voices. His support in establishing SDSC was invaluable. Before he became director of RSPacS, he had established a relationship with the Ford Foundation, and it was on Ford money that the Centre was initially set up. By the time Bob O'Neill came to request general university funding for the Centre, the Ford Foundation money was running out, but it was clear by then that SDSC had filled a need and was an academic success. Besides, there was no one equal to O'Neill with a grant application.

Next let me pay a tribute to the successive heads of the Centre. When Millar began, he had very little to go on, apart from the experiences of the other similar bodies that he had visited overseas. I had some knowledge of two of these — the Institute of War and Peace Studies at Columbia University (where I had had many talks with Bill Fox) and the Institute for Strategic Studies (ISS) in London (where I was friends with Alastair Buchan) — and I could see that neither pattern would exactly fit the needs of the Canberra situation. It was a remarkable achievement on Millar's part to make his conception of the Centre agreeable to university opinion in Canberra: here was something new, untried, vaguely open to the charge of militarism, and lacking in precedent.

---

1   L.T. Evans & J.C.B. Miller (eds), *Policy and Practice: Essays in Honour of Sir John Crawford* (Canberra: ANU, 1987).

Looking back, I can say that it greatly benefited the Centre that its first two heads had professional military experience. If either Millar or O'Neill had been solely academic in background, the credibility of SDSC, especially in military quarters, would have been much slighter. As it was, these men could not be ignored. General Sir John Wilton told me that when O'Neill went into academic life, Australia lost a future chief of the general staff; but did it matter? Instead of aspiring to what is sometimes an honorific position, O'Neill proved to be a wise, creative and forceful head of SDSC, someone who guided it into the waters that it commands now; and his subsequent career showed how right the decision had been to ask him to head the Centre.

When we moved to Des Ball as head, SDSC, as it were, took up a position of independence within ANU. Before that, it had been an advantage that the head was also a member of the Department of International Relations. In particular, during O'Neill's long period as head, it was worthwhile to combine the forces of the two entities, and for the head of International Relations — usually me — to lobby intensively for increases in resources for SDSC. I could also be brought out, like an ancient howitzer, to throw a shot at the Department of Defence if it was proving difficult. But those days are past — except, I hope, in less formal but no less intensive forms. The Centre now stands proudly on its own feet.

Let me also mention the influence of Hedley Bull upon the development of SDSC. Bull was joint head with me of International Relations from 1967 to 1977. As such, he had a great deal to do with the Centre. He supported it wholeheartedly — though there were moments when he and O'Neill were in dispute over the respective spheres of SDSC and International Relations. What Bull brought especially to the Centre — and what Millar, O'Neill and Ball have all brought in their separate ways — was making it a significant part of the worldwide network of institutions concerned with strategic studies. His reputation in this field was substantial. As the author of *The Control of the Arms Race* in 1961,[2] and as one of the founders of ISS, he was known in the United States, Britain and Europe. If he said the Centre was worthwhile, people believed him; and that is what he said.

---

2   Hedley Bull, *The Control of the Arms Race: Disarmament and Arms Control in the Missile Age* (New York: Frederick A. Praeger, 1961).

A summary of my thoughts about the establishment and progress of SDSC would go something like this: SDSC arose from a combination of the thinking of such people as Crawford, Millar and myself with the kindly assistance of the Ford Foundation; it has been fortunate in being guided by men of integrity, scholarship and shrewdness; the establishment owed much to the temper of the times, in which questions of strategy and defence had been either heavily professional or a political football; it gained something from overseas examples but developed very much as an indigenous enterprise; and it benefited greatly from successive directors of RSPacS — not only Crawford, but also Oskar Spate and Wang Gungwu and their successors. One can also say that the rise of the Centre occurred at a time when the impact of the social sciences on public policy was greater than in any previous period of Australian history. We now have more intelligent discussion of Australian defence policy in the newspapers — I exempt the odious medium of television — than ever before, and also of economic and social policy. This is a more thinking Australia than throughout the 1950s and 1960s; in that sense, SDSC has been part of a general movement towards more effective and more open discussion of policy than I grew up with.

To say that SDSC has been part of a general movement is in no way to diminish its achievement. It simply means that it caught the wave that applied to it, and rode it triumphantly to the beach. You will not find many other examples of academic bodies that have affected public policy and raised the level of public discussion to the same extent. We have here a remarkable academic development that has not just affected public policy in the sense of putting forward views on Australian defence that challenged established nostrums, but has also, through the research techniques of Ball, helped to undermine the childish and ineffectual secrecy that was so characteristic of Australian governments, partly because of its own protective carapace, and partly because of its being in thrall to British and American intelligence.

Let me now, in conclusion, say something about where strategic studies might be going, here and elsewhere. I have been little more than an observer of their development; seven weeks as an officers' barman in World War II hardly qualified me as a participant, and I never acquired either the background or the language of the strategists. But an onlooker sees some of the game; so I can perhaps venture a few ideas.

Strategic studies shares with certain other cross-disciplinary areas — such as women's studies, media studies, cultural studies, and criminology — the academic problem of not belonging directly to a particular traditional discipline. It also shares the problem (if it is a problem) of being policy-oriented. The first of these is mainly a career problem, the second a political one.

In academic terms, the difficulty inherent in cross-disciplinary studies is that, unless they grow rapidly and spread across the university scene at large, careers within them offer little hope of promotion, and the difficulty of finding careers within the conventional disciplines becomes greater. So far in Australia this difficulty has not been acute. The fact that, for most of its history, SDSC's graduate students were formally enrolled in the Department of International Relations meant that they could get teaching jobs under that heading; and indeed there was so much overlap between the two that this was good for both of them. Ball came up that way. Perhaps, as strategic studies becomes more specialised and moves away from the dominantly political element in international relations, there may be problems — and perhaps not. I am not trying to lay down any law, but am merely speculating.

The policy-orientation issue is inherent in strategic studies as such, since strategy is policy, defence is policy, and very little can be said on either that does not, or does not seem to, criticise or commend particular government policies. Even putting forward an alternative policy is to imply that the existing one may be no good. SDSC seems to me to have met this problem fairly and effectively. Successive heads have seen the need to get to know the right people, to achieve mutual trust with them, to be honest in telling them what is proposed and how it will be effected, and to involve them wherever practicable in the Centre's affairs. At the same time, SDSC has preserved a reputation as a non-partisan, informed and honourable source of comment on policy. This continued under Paul Dibb's leadership.

So far as strategic studies as a subject or discipline is (or are) concerned, I see a continuing future, if only because of changes in technology. We are not going to see many, if any, reductions in the number of sovereign states, and industry will keep on inventing new weapons and new ancillary means to make them effective. The Gulf War of 1990–91 made us all aware of the enormous advances in weaponry over previous wars; it showed us how a state can remain viable, even under the impact of the new weapons, and how contradictory and yet potentially dangerous can be the weaponry of a relatively small Third World country — especially if it has oil revenues or their equivalent.

I do not believe that the end of the Cold War will lessen the need for strategic studies, though it is true that it was largely the possession of nuclear weapons by the superpowers that provided the impetus for the discipline in the 1950s and 1960s. There will still be substantial military forces at the disposal of the great and major powers; there will be disturbances in the Middle East and the Balkans, often involving outside powers; there is still the possibility of confusion and even conflict in East Asia; and there is the permanent instability of Africa. There will continue to be conundrums such as we see in Fiji and Sri Lanka, where external military intervention may seem to some countries and even local politicians to be a solution to communal discord. There will be plenty to study.

I hope the studies will not become too narrowly specialised or too mathematical. War is one of the most complex and uncertain activities in which human beings engage. It involves science, technology, manpower, professionalism, economics, social change, and often the most intense forms of politics. Its outcome has rarely been predictable. Its avoidance involves diplomacy and luck. To see it as a whole — whether it is happening or being avoided or simply prepared for — requires various kinds of expertise. I hope they will all thrive within SDSC.

Finally, allow me to say how glad I am to have been associated with SDSC, if only in peripheral ways. ANU has had varying fortunes with centres and units; none has been more successful than this. Wisely and smoothly managed by Tom Millar, Bob O'Neill, Des Ball and Paul Dibb; with the benevolent efficiency of Jol Langtry; with a great many contacts abroad; with the cooperation of far-sighted people in government; this has been a model enterprise. Long may it flourish!

4

From Childhood to Maturity: The SDSC, 1972–82

Robert O'Neill

This essay was previously published in the 40th anniversary edition. It is reprinted here in its near original format.

Life at the Strategic and Defence Studies Centre (SDSC) in the early 1970s was not for the faint-hearted. One morning I turned the doorknob of Seminar Room A on the ground floor of the H.C. Coombs Building to find a room full of agitated people, some of whom were brandishing placards that called for the visiting seminar speaker, who was right behind me, to be tried as a war criminal. To underline the seriousness of the case in the view of those with the placards, there was a stout rope noose hanging from the facing wall, presumably in the interests of a swift execution of sentence. I took in this scene in a split second, wheeled my speaker about and we returned to my office. As any novice strategist could detect, this was no time to persist in the hope of having a quiet hearing for the speaker followed by a robust but regulated debate.

Our intended speaker that morning was Douglas Pike, a well-known American analyst of Viet Cong organisation, leadership and methods of operation. Once in the Central Intelligence Agency (CIA), he had relocated to academia. That link in his past was sufficient to attract the radicals and vitiate the prospects for an informed discussion on one of the most important issues of the Vietnam War — the nature of the communist movement there. I recount the episode as an illustration of the delicate way in which I had to conduct SDSC's activities while the Vietnam War was still a major issue on The Australian National University (ANU) campus. Despite having served in the Australian Army in Vietnam myself, and having a keen interest in seeing different methods applied by the Americans in that conflict to those conflicts they were following (to put it no more strongly), it was unwise to try to debate the war in an academic seminar room, at least until Australia's complete withdrawal was announced by the Labor Government of Gough Whitlam in late 1972.

Fortunately there were many other issues of a less inflammatory kind on which to focus, such as the changing balance of strategic interests and actors in the south-west Pacific. Unfortunately, the Centre by 1971 had very little by way of resources with which to sponsor research, publications and debate. I took it over from Tom Millar, the founding head of SDSC, in early 1971. Tom was appointed Director of the Australian Institute of International Affairs (AIIA) a year or more previously and was overloaded. I had not been keen to take on the headship of SDSC because, having recently come out of military service and then teaching in the Faculty of Military Studies at the Royal Military College, Duntroon, I desperately wanted to press ahead with my own research and writing. At that early stage of my career, I did not want to be saddled with the responsibilities of administration, the battle for financial support, and the planning and direction of conferences and seminar series that were part and parcel of the life of any head of an academic unit. Nor did I want to incur the opprobrium that would inevitably go with a public role in this field during the Vietnam War unless I turned into a radical opponent of the American and Australian part in that conflict.

When Hedley Bull, then head of the Department of International Relations in which I served, ran the question of my succeeding Millar by me in early 1970, I replied in the negative. Twenty-four hours later, after a good deal of soul searching and discussion with my wife Sally, I advised Bull that I would be willing to take on the SDSC headship. It would have been too selfish for me to have stuck to my initial preference, I thought, and the foundations of a serious analytical capacity in the field of international security at ANU would have been weakened had I not been willing to take the burden of running SDSC off Millar's shoulders.

Thus the die encasing my academic career was cast. And not a bad cast it was, leading me on to 20 years of international service in positions of this kind after 11 and a half years as head of SDSC. Nonetheless, the decision cost me in that it was difficult to carve out the time in my schedule for research and writing on a personal project, especially as I had already agreed to write the official history of Australia's role in the Korean War, partly to break the logjam in Australian war history that was blocking the path to a full analysis of the Vietnam War and its lessons for the future. My view of my future career path was to return to writing the kind of book that had resulted from my doctoral thesis at Oxford, *The German Army and the Nazi Party*, and thereby become better known internationally, and maybe gain a significant position in a major university abroad. Leading SDSC seemed to take me in another direction altogether, which would make me better known in Australia but would probably lead me to being typecast as a regional rather than a global scholar.

When I took over the headship of SDSC from Millar, the Ford Foundation grant on which the Centre was founded was almost exhausted. SDSC had been established on one of the worst of bases for an academic research unit: it had no funding from ANU other than providing for my own salary, which was a charge on the Department of International Relations in which I had a tenured position as a senior fellow. The Ford Foundation grant provided for a secretary and a research assistant, but this would not continue for long. There were no research positions in SDSC and no administrative assistance other than the above two posts for running the conferences and seminar series that were essential to a centre's visibility. I felt for a while as though I was imprisoned inside a hollow pumpkin with a very thin shell.

Some welcome relief came from the Business Manager of the Research School of Pacific Studies (RSPacS), Peter Grimshaw, who believed strongly in the Centre's work. He found ways in which some unexpended funds in the RSPacS budget might be moved to support SDSC, provided that the Director of the School, Oskar Spate, and the School Faculty Board approved. It was partly my task to persuade Spate that this would be a wise use of School funds. He and the board assented and the long-term consequence of that decision was that SDSC had an assurance that it would always have some secretarial and research assistance provided by ANU.

It was, of course, more difficult to obtain funding for academic research posts because these were far larger budgetary items. It did not help me that RSPacS had recently established a Contemporary China Centre because (despite being a later arrival than SDSC) it was seen as offering greater opportunities for RSPacS and ANU to raise their profiles, not least with the Whitlam government when it came into office in 1972. I never tried to use the argument of having stood longer in the queue to gain ANU financial support for SDSC: it would have been very counterproductive. Yet that did not stop me from feeling somewhat lonely and more than a little disappointed at the prospects I could see ahead for SDSC for 1973 and beyond; I wondered if it would fail on my watch. No research meant no scholarly impact. Without impact in the 'dog-eat-dog' world of academic departments and centres, SDSC would soon be 'dead meat'.

The fortunes of politics opened up an avenue of hope to me in late 1972. During 1971 and 1972 I had come to know Lance Barnard, the Deputy Leader of the Australian Labor Party and its Defence spokesman. He had spoken to a Centre seminar and privately I had been able to answer some of his questions concerning the Australian Defence Force (ADF) and its equipment needs. He certainly had my respect as an alternative minister for defence, and I felt that he saw SDSC as something that might be developed further in the interests of a more open and informed national debate on security issues. Furthermore, his principal assistant, Brian Toohey, later to become well-known as a journalist and editor, lived across the road from me. Once the electoral outcome was known in late 1972 and the two-man Whitlam–Barnard Government was installed in office, I began talking to Toohey about the possibility of the Defence Minister funding two non-tenured posts in SDSC — a research fellowship and a senior research fellowship.

I knew that there would be controversy within ANU about any direct government funding of defence research in academia, but as it was a Labor Government that would be providing the funds, the academic hubbub soon died away. Barnard, for his part, did not grant my request at once. In a normal bureaucratic way, the proposal was looked at closely within the Defence Department and at a senior level in the armed services. It received more support than opposition, due also to help from Bruce Miller, Bull and Millar, both within ANU and in negotiations with the Defence Minister. The finance was granted and was accepted by ANU and RSPacS and, by 1974, SDSC had an entirely different prospect to the bleak outlook of 1972.

During 1971 and 1972 I had been developing a parallel track to SDSC. Because of the political and security sensitivity of their work, many people in the armed services and Defence (both civilians and military officers) were reluctant to participate in any kind of public discussion of their field. Yet they had much to contribute without infringing official regulations. The highly charged atmosphere of the Vietnam War in the early 1970s was a formidable disincentive to people within the defence establishment who might want to take part in SDSC seminars and conferences. Yet, I knew from experience in Canberra and London in the 1960s that there were some very bright minds in government service who had much to offer our work in SDSC, not least by criticism of the ideas we were beginning to develop. Indeed, without their participation, we stood to suffer in terms of focusing on the right issues and having access to relevant unclassified information and experience.

Each of the six states of Australia had a United Services Institute (USI) that fostered professional debate, largely by persons inside the defence establishment or those who had retired from it. This was a true colonial legacy: the Australian states may have federated, but the USIs had not. So I set about enlisting support from some of the brighter military officers I knew at the level of colonel or thereabouts, together with civilian members of Defence around the grade of assistant secretary and below, for the establishment of a discussion group that would constitute the USI of the Australian Capital Territory (ACT).

The founding of the USI of the ACT broke the ice that previously shut in some of the better brains of the defence establishment and we soon had a flourishing new organisation holding lectures and conferences. We were fortunate in 1971 that Michael Howard, one of Britain's leading strategic thinkers, was visiting ANU and was able to contribute the opening lecture for the USI of the ACT. The group continued to build up momentum and finally fulfilled its long-intended role of being the keystone in the arch of a federated Royal United Services Institute

(RUSI) of Australia, under which all the state and territory USIs were brought together. RUSI of Australia is still functioning as a valuable forum for the development and criticism of defence thinking, and its establishment has further helped to reduce the barriers impeding the free exchange of ideas between the academic and government sectors of the defence community in Canberra.

Returning to the story of SDSC, substance was soon given to my hopes for its future research capacity because, in the first few months of 1974, using the Defence Minister's grant, we were able to appoint two persons who were to make major contributions to SDSC's reputation: distinguished journalist Peter Hastings, and recent doctoral graduate Desmond Ball. Hastings, a generation senior to Ball, was appointed to the senior research fellowship and Ball to the research fellowship. Ball had produced a bold and brilliantly prepared thesis under Bull's supervision on the strategic nuclear policies of the administration of US President John F. Kennedy only a decade after the decisions he was analysing had been made. He demonstrated amazing capacities to unearth crucial and sensitive evidence. He quickly established credibility with well-known American scholars and senior figures in the Kennedy administration and, indeed, had a far bigger reputation in the United States than in Australia.

At ANU, Ball's image was more that of a radical student activist, opposed to the Vietnam War and critical of many of the ways by which Australia had been governed by the coalition parties over the past generation. When Sir Arthur Tange, then secretary and permanent head of the Defence Department, discovered that Ball had been appointed to this post, he vented his displeasure on me. Tange believed, wrongly in my view, that Ball was a malicious troublemaker, who was more likely to damage Australian security than to strengthen it. Over the following three years, I sat in Tange's office on several occasions listening to his complaints about critical analyses that Ball had written of the way in which defence-policy decisions were made in Australia, and of the problems that some aspects of US nuclear weapons policies, and their installations in central Australia, posed for Australia. There is no doubt that Ball was arguing against the government's line of policy, but that is part of a scholar's job. He gave us insights, based on a huge body of research, that we needed to have. Fortunately, Ball, through his personal qualities and his standing in SDSC, was soon invited by the armed service staff colleges and similar institutions to lecture, and he made his mark independently. His career took off, aided by the regard in which he was held by leading American scholars and by his remarkable access to persons in the defence establishment in the United States.

As a promising young scholar in a huge, complex and crucially important field — nuclear strategy — I thought ANU should try to keep Ball for a longer period than the 3–5 years normally associated with a research fellowship. This meant appointing him to a tenured fellowship, such as the post that I had within International Relations. For some of my colleagues, the idea of a tenured post being offered upon the establishment of a centre was unacceptable. Centres were meant to be able to be folded up and put away at short notice when the money or academic interest ran out. I believed that the work of SDSC was too fundamentally important as an academic endeavour for the collapsible model to govern its development forever, so I began a campaign for a tenured post in SDSC, specifically with Ball in mind as its first occupant.

Again Ball helped his own cause powerfully by going off as a research associate to the International Institute for Strategic Studies (IISS) in London for a year (1979–80). This was the pre-eminent international body in our field and it was a feather in Ball's cap to be invited there. While in London, he produced one of the best research papers that IISS has ever published — on the feasibility of successfully conducting a limited nuclear war. Ball's analysis argued that the network of sensors and communications on which control of a nuclear engagement depended was too vulnerable to survive long should hostilities occur. Therefore, the many attempts that had been made by politicians, officials, military leaders and other scholars to build public and professional confidence in the usability of certain kinds of nuclear weapon in a major conflict had little substance.

This paper hit the headlines — both globally and at ANU. It was thus easier for me to win the battle on tenured posts for centres once senior academics knew that there was a real danger of Ball being attracted away to a leading US university. Ball came back from London and was appointed to a fellowship in 1980. The continuing quality of his work led to his promotion through the two higher grades to become a full professor a few years later. SDSC has been splendidly strengthened by Des Ball for over 30 years since. He followed Millar as its head in March 1984 but, realising that research rather than administration was his forte, was pleased to be able to pass the headship on to Paul Dibb in 1991. Ball has been wonderfully hard-working, perceptive, enthusiastic for his and the Centre's work and warm in his relations with colleagues, supporting staff and students. For most of his time in working with myself in SDSC, he was its deputy head and was splendid in that role. I could always count on him for loyal and energetic support, and a flow of fresh, high-quality ideas relating to the future work, development and financial support of SDSC. We have remained in regular contact since my departure from SDSC in 1982 and I have always regarded him as a personal friend.

Hastings also made a strong impact on the Centre, despite being with SDSC for only three years. He needed to return to his profession as a journalist and I had no hope of arguing successfully for a second tenured post for him. Yet, in the time that he was with the Centre, he taught us all much about the region to our north in which he was specialised. He wrote on the implications for the region of its rapid political, economic and social development, and on Australian policies for meeting the challenges coming in our direction. He was an immensely witty man, with a huge range of personal connections in government and beyond — in Canberra, Sydney, Indonesia and Papua New Guinea. He was a connoisseur of many of the best things life has to offer and was great company on a field trip. We did two together — one over north-western Australia for several weeks and the other over north-eastern Australia. With his journalist's influence, he was able to get the use of a light aircraft, which made a huge difference to what we could see of the two regions. We were familiarising ourselves with the defence problems of northern Australia — issues that had been ignored since the dark days of the Japanese threat during World War II, but which were demanding fresh attention with the cessation of the time-honoured policy of 'forward defence' following the Vietnam War.

It was a rich personal experience to work closely with Hastings over these years. He shared my interest in modern history and we had much to talk about and learn from each other. One lasting legacy he left me was personal access to a whole range of senior people in both government and the diplomatic service who were his friends and contemporaries, but 15 years or so senior to myself. From those contacts I learned much about the scepticism and detachment that senior bureaucrats have towards the political masters they serve every day.

In 1974 and 1975, the SDSC program of activities focused primarily on the strategic nuclear balance in the world at large. We held a conference on this theme in each of these two years. The first was conducted with the initial five-year review of the nuclear Non-Proliferation Treaty in mind and, two months before the conference, the appositeness of our judgement was underlined by the Indian nuclear test on 18 May 1974. As with most conferences, it was a test not only of the speakers' abilities to produce good papers but also of the Centre's standing in terms of who accepted the invitation to attend and who else applied to participate. We had encouraging results on both scores. The authors (Arthur Burns, Ball, Harry Gelber, Geoff Jukes, Peter King and Jim Richardson) gave us good analyses from differing perspectives and we decided to publish the papers as a book. This aim was more difficult to achieve than it might have been because Australian publishers were not greatly interested in defence topics in 1974, so we had to use some of SDSC's small discretionary funds to produce the volume ourselves. The Centre's secretary Jenny Martyn typed up the whole volume, while our research assistant Jolika Tie (later Jolika Hastings after marrying Peter) checked all the footnotes and undertook the copyediting.

I was the overall editor and contributed an introduction that, having just re-read it after a break of 32 years, I am still happy to put my name to. Some thought it overly pessimistic at the time, but I was not wrong in pointing to the long-term significance of India's 'peaceful' nuclear explosion both for India and for those thinking of developing a nuclear option in other countries. The non-proliferation regime looks even weaker now than it did to me in 1974. The volume remains available for purchase second-hand on the Internet, so someone must think it is still worth the advertising space.

This experience of self-publication opened up a new avenue of opportunities. With highly productive colleagues and topical subjects under analysis, SDSC had real promise for developing its own publications program. Our publications developed gradually through the mid and late 1970s from a small series of Canberra Papers on Strategy and Defence to a much larger series of SDSC Working

Papers, which were produced from typescripts, printed on an office printer, and then placed between identical stapled covers that differed only in a window displaying the title and author. While we retained the Canberra Papers series, we decreased the quality of the format, thereby saving us money, and put a major effort into marketing both these and the Working Papers. The production rate increased and print runs lengthened. We also began producing books at the rate of two or three a year, including a handbook of data and analysis, *Australia's Defence Capacity*, beginning in 1972. This volume underpinned the public debate with a basis of accurate information ranging from Australia's diplomatic assets and liabilities, such as treaties and other commitments, through to defence forces, bases and equipment. The publications program was largely self-sustaining, and it was led by a third newcomer to the Centre's staff, Colonel Jol Langtry.

Langtry arrived at SDSC in 1976 as a result of general agreement by the RSPacS Director, Wang Gungwu, and his colleagues that SDSC needed a staff member to relieve me of some of the burdens of conference and seminar series organisation and also direct the burgeoning publications program. Grimshaw, also played a major role in this appointment, both in finding the money needed for the new post and giving me some leads on how to play my hand in the internal competition for resources in RSPacS.

Langtry was a huge asset. He was one of those rare infantry officers who had both a university degree and had won the prized Distinguished Conduct Medal as a sergeant in the New Guinea campaign in World War II. He came to SDSC in an administrative capacity, but it was soon abundantly clear that he could conduct research and write at the standards required for a university centre, so he became a close colleague in every possible way. He made a huge difference to my workload and he opened up a presence for SDSC in many new ways. He was independently minded, strong, capable and humble. He fitted into a small, informal academic research unit with great ease and everyone in the Centre valued him highly. He soon teamed with Ball and Ross Babbage, then a doctoral student in International Relations (centres were not permitted to have students in the 1970s, so we supervised those who were specialising in strategic studies under the aegis of International Relations), to produce joint works with them, as well as publishing solely under his own name. Langtry also recruited a team of support staff to help in the publications process.

The computer (as we now know it) was just coming into vogue and he was able to have an office suitably equipped with terminals and staffed by proficient operators. SDSC's publishing capacity became formidable. Indeed, it was more advanced than that of IISS, as I was to discover when I arrived in London in August 1982. What I learned from Langtry's program stood me in good stead for initiating a major modernisation of IISS's capabilities in this direction.

Again, due to Langtry's administrative skills and initiative, we were able to mount more influential conferences that ran over two days and brought up to 300 specialists together. From 1974, SDSC held one major conference a year and sometimes two. We mounted a special effort to build up the Centre's mailing list, which linked security specialists in the Australian academic world, interested politicians, public servants, armed service personnel, journalists, business people and industrialists, increasing it to over 600. We produced a quarterly newsletter, which the Centre still finds useful to publish (another idea I took to IISS, where its version still appears regularly). Publication sales, and seminar and conference attendance all rose. SDSC became an obvious and very active hub for anyone in Australia interested in serious discussion on, and research into, national and international security issues.

The year 1976 was notable for launching research on our second area of specialisation: the defence of Australia. At the conclusion of our 1975 conference on the strategic nuclear balance, several of us adjourned to the bar in University House at ANU and the suggestion arose from Babbage that we might focus our next conference on how Australia, in the era after that of forward defence, might be secured against international pressures or aggression. The virtual collapse of forward defence after the US and allied defeat in Vietnam left Australia without any readily comprehensible strategic policy. This was a major opportunity for a group of academic specialists to open up the topic, try to identify the main questions and problems that needed to be addressed, and then begin to formulate some answers to the questions and solutions for the problems. It was clearly an ambitious undertaking to hold a conference on this theme, but we had an excellent team in place with which to tackle it.

Babbage developed some interesting ideas on this subject in his Master's thesis at the University of Sydney (which I examined). He built effectively on this foundation in his doctoral work at ANU and was in an excellent position to make an impact in public debate. Apart from Ball and Hastings, we were able to call on two international specialists with appropriate expertise: Sven Hellman from Stockholm, who was a specialist in defence planning with an emphasis on self-reliance; and James Digby from California, who was a specialist in new weapons technology and its impact for the strengthening of defensive capacities. By October 1976, when this conference was held, we also had the previous Australian defence minister, Bill Morrison, working as a visiting fellow in the Centre. He joined the team of speakers by offering some thoughts on the impact of the recent basic re-organisation of the Department of Defence in Canberra and the role of the minister in defence planning. Some 120 people came to hear the seven of us and join in debate, and the result was a path-breaking book published by SDSC the following year, *The Defence of Australia: Fundamental New Aspects*. It helped to set the course for the policy debates on Australian defence by influencing the ideas put forward in the Australian Government's Defence White Paper of 1976 — the first to grapple with Australia's defence strategy after the withdrawal from Vietnam. The book, and the subsequent work of its authors, continued to have an impact over the next several years, leading up to the report prepared by Dibb (a former senior member of Defence and then a special ministerial advisor, later the fourth head of SDSC) in 1986 on Australia's defence policy in the broad, and then the Australian Government's Defence White Paper produced in 1987.

SDSC followed this conference with a second the following month, November 1976, on 'The Future of Tactical Airpower in the Defence of Australia'. The major procurement issue before the Australian Government then was what aircraft should be purchased to replace the Mirage III–0. As Ball wrote in the introduction to the resulting book, we were not attempting to 'pick a fighter for Australia' in this conference, but rather 'to discuss the general philosophical, technological, strategic and industrial questions relating to this decision'. Langtry joined the team as a speaker and paper author. Other members were Ball, Babbage, Kevin Foley, David Rees, Peter Smith and myself. Ball edited the book of the conference, *The Future of Tactical Airpower in the Defence of Australia* — another SDSC publication that met a strong market demand. Although we did not attempt to choose a particular aircraft, when the government's final choice rested between the General Dynamics F–16 and the MacDonnell Douglas F/A–18, our arguments tended to favour the latter. We felt some satisfaction when the F/A–18 was selected because it fitted our criteria more closely.

I shall not expand on the work undertaken in SDSC relating to the defence of Australia in the late 1970s and early 1980s for reasons of space. Suffice it for me to say that research and discussion in this area remained a principal activity of the Centre. Its sustained high profile in the media, and the degree of interest shown by Defence, politicians with defence interests, the Canberra diplomatic corps and the academic community generally gave us the feeling that our work was playing an important role in facilitating the development of rational policy on major issues of Australian defence policy.

The Centre's work on regional security problems was a third major field of activity. During the Vietnam War, it was inviting trouble to attempt to work on either Vietnam or South-East Asia from a security perspective, so we focused on the newly independent states of the South Pacific, which, by 1975, included Papua New Guinea. Hastings was our principal expert in this field and, as the Vietnam War receded, we were able to focus more directly on South-East Asia — not least East Timor. José Ramos-Horta (later Foreign Minister of East Timor) visited a few times during the late 1970s and spoke at seminars on issues that were to become much more serious — even tragic — in the late 1990s. Our conference program in the late 1970s and early 1980s focused heavily on regional issues embracing South Asia, South-East Asia, and the Indian Ocean.

This work was strengthened considerably through the financial support of the Ford Foundation. After its initial grant (which was vital to the establishment of SDSC in 1966), the Ford Foundation focused on other areas of academic endeavour. During 1974, Enid Schoettle, head of the international security program at the foundation, contacted Bull and myself with a view to discuss a wide project of international academic cooperation in which International Relations and SDSC might play leading roles. She came to Canberra and the net result was funding for a Master's degree program within International Relations for outstanding students from South and South-East Asia and for a visiting fellows program in SDSC directed towards the promotion of regional stability by arms control and sensible, non-provocative defence policies. The grant also supported a major international conference to be run by SDSC each year, and travel by ANU staff members both to select candidates for graduate scholarships and visiting fellowships and to give lectures at the universities we visited in South and South-East Asia.

The Ford program came into effect in 1975 and was extended for several years beyond 1978 — the initial period of the grant. It enabled us to bring to Canberra generally two visiting fellows a year, sometimes more, who stayed for six months, contributed their perspectives to our work, learned what they could from us and contributed a major piece of research for inclusion in our publications program, generally as a Canberra Paper. The Ford program also gave us an opportunity to extend our network of contacts, inputs and influence from Australia to the wider region. SDSC became known as a regional rather than a national centre and this helped further in gaining us attention in Europe, North America, Russia and even in the People's Republic of China. The Ford Foundation, in addition to its grant to ANU, sponsored annual gatherings of the institutions it supported around the world in the field of international security, and these meetings were another great opportunity for discussing our work with our peers in other parts of the world and leaders in the field from universities such as Harvard, UCLA, Columbia, Tokyo, and Jawaharlal Nehru in New Delhi, the London School of Economics and Political Science, and the Graduate Institute of International Studies at Geneva. For the next 20 years and beyond, senior scholars and some government officials in our field who had been through the Ford program in SDSC were to be found in important positions in South and South-East Asia. I am still in touch with several of them.

The nature of the work undertaken in SDSC, especially but not only that of the regional security program, attracted the attention of many members of the Canberra diplomatic corps. In the mid-1970s, there were some 70 foreign diplomatic missions in Canberra, ranging from those of Australia's closest friends and partners — such as the United Kingdom, the United States, Canada and New Zealand — through to those of states that were far from close friends of Australia in those days of the Cold War — such as the Soviet Union, China, East Germany and the countries of Eastern Europe. SDSC developed close links with other embassies such as those of Sweden, Germany, France, Belgium, Italy,

Israel, Egypt, Japan, South Korea, and those of nations from South-East Asia. We had little time to be charitable and our relations with diplomatic missions turned on their utility to SDSC. Many of them were helpful in facilitating the travel arrangements of SDSC members to their countries, fostering relations with their relevant academic and governmental institutions, funding visits from their scholars as conference speakers in Canberra, and keeping us informed of their government's policies, their internal national debates and their views on the wider issues that concerned us all.

Interesting social interactions often went hand in hand with these professional diplomatic connections — these were not always pleasant or easy, especially when one knew that one was being entertained by, for example, the head of the East German intelligence cell in Canberra. Generally we had a vigorous relationship with the members of the Soviet Embassy, who not only contested our views on many international issues but also directed a major intelligence gathering effort in our direction. Yet, in the cause of international peace and brotherhood, we gave as good as we received by way of debate and it was a relief to discover that many of our challengers had an excellent sense of humour and profound scepticism towards their own governments and systems of political organisation.

We thought it particularly valuable to have (as a visiting fellow in RSPacS) the occasional Soviet specialist on Asia and the Pacific. They tested our thinking and gave us a deeper understanding into the reasons for policy differences, and perhaps our free-wheeling modus operandi had a subversive effect on the power of the Communist system over them. We had to be on our guard of course.

One morning in the late 1970s the *Canberra Times* ran an article on its front page stating that the space tracking station at Tidbinbilla, just outside Canberra, had acquired a new defence function. Not long after I sat down at my desk that morning, in bustled a Russian visitor in a state of excitement about the report. It seemed to confirm his view of America's power over Australia. I thought the report was wrong. There would have been much more evident security barriers around Tidbinbilla had it been true. As it was still possible for visitors to be guided over the tracking station, I suggested that next Sunday our guest might accompany my wife Sally, our daughters and myself on a picnic in the vicinity of the tracking station, after which we could visit it. Our Russian visitor accepted the invitation; we had a pleasant lunch on a creek bank nearby in beautiful weather and then walked up to the tracking station itself. We had ready admission, and took a guided tour. My guest remained silent. He had brought his camera with him and when we were out in the grounds again after visiting every building on the site, he asked if he could take some photographs. I said, 'Certainly — photograph everything you want!' A look of puzzlement and disappointment came over his face. He collected his thoughts for a minute, looked hard at me and then said, 'Oh, I don't think I will bother thank you.' The power of Washington seemed to have receded in his thinking that day.

The staff of the Soviet embassy in Canberra often included some lively people — none more so than Counsellor Igor Saprykin. I met him at a diplomatic reception shortly after his arrival, which coincided with a debate in the national media about the imperilled future of the red kangaroo. Saprykin wore a fine, dark-blue, pinstriped suit and a dark-blue silk tie with a red kangaroo embroidered centrally on it. I made an opening remark about the tie and he replied, 'Yes, don't you know this is my "save the kangaroo" tie — better red than dead!' In view of the then current controversy about the mortality rate of the red kangaroo in Western Australia, I gave him four runs off my opening ball. For the next three years we were to have some lively conversations. A little later, when one of the parliamentary committees produced a report strongly critical of human rights for those in the Soviet Union who were active in the arts, he was asked by a journalist for his comment on the committee's report: 'A pain in the arts!' was his verdict of dismissal.

For most of the 1970s and early 1980s, the Centre's small support staff carried a major workload. We had a research assistant who was in effect also the SDSC librarian, and a secretary who took all the dictation of correspondence in that pre-computer era, typed all the academic and administrative papers, answered the phone and received visitors. Our secretaries, in order of their tenure of the post during the 1970s, were Margaret Kewley, Jennifer Martyn, Anne Buller and Billie Dalrymple. The research assistants were Robyn MacLean, Jolika Tie, Anna Booth and Jann Little. They always had more to do than could be covered in an eight-hour working day and when the pace was particularly hectic — because of the coming of a major conference, the imminence of a publication deadline or an international crisis that required Centre academic staff to figure prominently in media commentary — they could always be counted on to put in a special effort whether they were paid for it or not — and usually they were not. With Langtry assisting, from 1976 the Centre had a very strong and well-knit team. They were highly intelligent and made sure that Hastings, Ball and I were well-equipped with information and recent commentary — especially when we were about to intervene in a media debate or appear before a parliamentary committee. They kept our publications up to standard in terms of readability, proofreading and accuracy of sources. They coped with a wide range of duties, ranging from showing visiting fellows to their living accommodation and how to use SDSC facilities through to arranging international air travel for Centre staff and conference speakers.

SDSC office accommodation was made by an allocation from International Relations of rooms on the middle floor of the H.C. Coombs Building. Bruce Miller, the department head, made available a consecutive run of rooms, including a large reading room for the research collection and in which visitors could work on our collection of books, journals, news clippings and other materials. Because of the pressure on accommodation in the Coombs, SDSC had just sufficient rooms for the academic and support staff to have one each, except for the research assistant who had to endure the lack of privacy associated with a desk in the reading room. My own office, as head of SDSC, was never more than the standard room designed for graduate students, support staff or junior academic staff; but at least our accommodation was in the heart of the building, right alongside International Relations. We could have moved away to more spacious premises in the further reaches of the campus, but I chose to stay close to the centre of things for a host of reasons, including avoiding marginalisation, demonstrating our closeness to International Relations in more ways than one, and remaining abreast of the scraps of useful political and administrative intelligence that floated around the busy corridors of the Coombs.

Once Hastings had completed his tenure as the Department of Defence-sponsored senior research fellow, his post was held by Philip Towle, a British political scientist who had a background in arms control and the Foreign Office. Towle was with SDSC until the early 1980s, when he was awarded a fellowship at Cambridge. With Ian Clark, a former graduate student in International Relations, Towle was to head the international relations program at Cambridge for some 20 years. Cambridge did not receive its first chair in this field until the late 1990s and, until then, Towle and Clark did virtually the whole of the teaching in this field there — a heavy workload due to the success of their Master's degree course.

SDSC junior posts were then occupied by persons who came for shorter periods. They included Ron Huisken in 1976–77 and Don McMillen and Paul Keal in 1981–83. Defence Minister Jim Killen responded favourably to suggestions from several quarters that the armed services should be able to give outstandingly able officers in mid-career an opportunity to conduct some professionally related research in an academic environment. Thus was founded the Australian Defence Fellowships program, which supported usually one or two officers in research projects at the Centre. As these fellowships were tenable in any Australian university, not all those selected came to SDSC. Nonetheless, by the time of my departure from the Centre in 1982, there were sometimes a dozen persons working there — a far cry from what had been the case a decade earlier. My successors have been able to build on this foundation to create and sustain a major academic institution. I think it has been many years since any head of the Centre has needed to worry about whether it would be in existence in one or two years' time!

As SDSC developed a critical mass of expertise and became busily engaged in contact with many defence-related organisations (both in Australia and around the world), we needed to think more about our social life together. The mix of personalities in the Centre was generally a very compatible one, and we often held lunch gatherings in the gardens of University House, enabling someone, say, who had been abroad at a conference or on research fieldwork, to bring the rest of us up to date with what they had seen and heard. A noteworthy annual event was a long summer weekend in one of the ski lodges on Mount Kosciuszko. Langtry was able to reserve the army ski lodge there for a few days at a time, which gave us a great opportunity to enjoy some relaxed time together in a beautiful place. The blooming of wildflowers in early February was wonderful to see and families, especially children, had some memorable times in each other's company. The frequent presence of visiting fellows not only from South and South-East Asia but also from Europe, North America, the Soviet Union and Japan, afforded many occasions for offering them some Australian hospitality in Canberra, leading to family connections that have lasted for several decades.

During the 1970s, there was an increase in the number of Australian journalists who took a serious interest in international security issues. Early in the decade, the predominant flavour of press comment on such matters was criticism either of the conduct of the Vietnam War or of American policies elsewhere. It was hard to get journalists interested in the wider aspects of regional security that were then our bread and butter. Ball was successful in finding media coverage for his work on decision-making in Defence and the higher reaches of government. Some of his writings were newsworthy in themselves and much of it offered unique perspectives of a kind that were readily available in the United States and Britain, while relative darkness reigned in Australia. Hastings maintained a high public profile by continuing to write for the *Sydney Morning Herald* while he was in SDSC. Gradually, more serious journalists from the print, radio and television media became interested in our work and, by the late 1970s, SDSC had become established as the first point of reference for journalists in Australia seeking comment on world events and government decisions of defence significance.

While many academics in the early 1970s tended to look askance at those few of their colleagues who made frequent media appearances, the climate of opinion was changing. Universities were finally beginning to feel the effects of government cuts in their funding, and alert vice-chancellors — such as Anthony Low of ANU — were keenly aware of the value of a significant media presence. During his years in office (1975–82), he gave me steady encouragement to sustain this activity and I think other benefits flowed to the Centre, enabling us to become less dependent on external funding and more reliant on ANU.

Another major part of our work was service to those parliamentary and other official committees and boards of inquiry that had oversight of international security issues. The body we came to know best was the Joint Parliamentary Committee on Foreign Affairs and Defence, and one or more SDSC members appeared as an expert witness in virtually every set of hearings that this committee held from the mid-1970s onwards. Appearance before a committee involved preparing an initial paper for submission, which was circulated to committee members and especially to their staff and, on that basis, the formal exchanges of the hearings themselves began. It was pleasing to find in the work of these committees that the more combative side of political competition was cast aside and to see that committee members were interested much more in getting at the truth and weighting professional opinions than in putting down rivals from opposing parties. In these hearings one could feel sympathy for those carrying some of the burdens of government as they sought genuinely, and under pressure of time and other commitments, to find the best policy solutions to the problems that Australia was facing in our field.

One major commitment we undertook for Defence was improvement of the courses conducted at the Joint Services Staff College (JSSC). This institution was established in Canberra around 1971 for the higher professional education of officers from the three armed services at the stage in their careers when they would be working increasingly in operations involving two or three armed services than within the confines of a single-service environment. The initial curriculum made some useful strides in this direction, but it fell a long way short of what Tange wanted in officers who were going to serve in senior policy-related positions, both within his department and at higher headquarters in the armed services themselves. In 1975 he established a committee chaired by ANU Deputy Vice-Chancellor Noel Dunbar to overhaul the curriculum and methods of teaching at JSSC, and service on this body took a substantial amount of my time over the following year. I then became academic adviser to JSSC until my departure in 1982. Ball, Langtry and Babbage also made major contributions to the work of JSSC.

As SDSC became more widely known, Centre members — especially Ball and myself — had increasing opportunities to take part in conferences and joint projects conducted by other institutions around the world, especially IISS in London. Bull had been a member of its council since the 1960s when he was based in London. I followed him onto the council in 1977, and Ball became a research associate there in 1979. We were also engaged in work with leading American institutions and with a wide variety of regional centres of research in Japan, South Korea, India, Pakistan and South-East Asia. The Ford Foundation continued its support of our outreach work, enabling us both to bring scholars and graduate students to ANU and to make a direct contribution to the work of their parent universities abroad.

The Australian Government and foreign governments supported us by paying for conference participants to come to Canberra and by funding our travel to see something of the work conducted in our field in other countries. Attendance at the SDSC main annual conference climbed to 300 and we were able to interest commercial publishers in producing books based on these conferences, which I or another member of the Centre's staff would edit.

All this activity took a toll on the amount of time and effort I could put into my own research and writing. In some ways my concerns about the negative impact that the headship of SDSC would have on my personal output were confirmed. With major effort, however, I continued work on the history of Australia's role in the Korean War (1950–53) and completed the second and final volume in 1982. This work took more time than I had imagined at the outset, not because of the military operations of the war (which were complex enough) but rather because the conflict had resulted in a transformation of Australia's foreign policy. The main feature of this period was the conclusion of the ANZUS Alliance, which Foreign Minister Percy Spender achieved against the odds — including Secretary of State Dean Acheson's personal reluctance, the opposition of the US Joint Chiefs of Staff and the scepticism of Prime Minister Robert Menzies. Other themes of the period were Australia's growing disengagement from Britain as a protector, the formation of regional links in South-East Asia, the delicate handling of China so that an enemy in limited war might be turned into a major partner in trade, the transformation of Japan from erstwhile enemy into Australia's principal trading partner, and the battles of Australian national politics of the early Cold War period. I had unrestricted access to government papers. It was a great challenge. I decided to address it by focusing one volume on the broad policy issues of the period and a second on the combat operations of the Korean War. I had hoped in 1969, when I took on the commitment, that I could write the history in four years. It took me 12. I had not, of course, reckoned on the other things that I would be doing, because SDSC was then not on my radar screen as a personal responsibility.

It was extremely fortunate for me that, by September 1981, I had completed the drafting of the second volume of the Korean War history (the first had been published earlier that year). The then director of IISS, Christoph Bertram, asked me that month if I would consider being a candidate for his position, which he planned to vacate in August 1982. This was another great opportunity as well as a huge challenge. I said 'yes' and, after six months of a detailed international assessment process, I was offered the position out of a shortlist of four. I was the only non-European on the shortlist and all previous four directors of IISS had been Europeans. Clearly we had done something in SDSC to overcome the barrier of remoteness from the principal centres of debate in the field of strategic studies, but also I felt that the IISS council's confidence in offering me the directorship rested substantially on what everyone in SDSC had accomplished over the previous decade. I was not new to the game of running a research institute and my colleagues had produced notable work that was regarded as of international calibre.

I took some satisfaction from the way that the SDSC's reputation had been transformed during the 1970s. In 1971 it was seen as a potential source of trouble within RSPacS and ANU because of the controversy surrounding Australia's part in the Vietnam War. The Centre had scarcely any funds and few friends. By the early 1980s, it was a strong card in the hands of both RSPacS and ANU, attracting money, high-calibre academic members and visiting fellows. SDSC was a focal point of the national debate on security issues and a well-recognised entity within the international network of research institutions in the field of strategic studies.

A good slice of the credit for this result belongs to a number of key external friends of the Centre. Miller and Bull, as alternate heads of International Relations, were constant supporters in provision of advice, use of their contacts nationally and internationally, contributions to SDSC's substantive work and, above all, assistance in building good relations and a strong base in the shifting sands of politics at ANU. I owe a major debt to Wang Gungwu (Director of RSPacS from 1975) for his ready assistance, and to his successor Gerard Ward. Anthony Low, who preceded Wang Gungwu for two years before becoming Vice-Chancellor, was another strong supporter to whom my thanks are due. The affairs of SDSC were also overseen by an Advisory Committee (advisory to the director of RSPacS), which consisted of senior members of ANU who had relevant interests. They were drawn mostly from Pacific Studies, but other schools were also represented and the committee served to integrate SDSC more fully into ANU as a whole, rather than remaining purely an element of Pacific Studies. Grimshaw always gave SDSC his support — particularly through a flow of suggestions for gaining the resources that were vital to the Centre's existence and growth throughout that decade. He did not have discretionary power over any RSPacS funds that were relevant to SDSC needs, but he could, and did, give me advice that led to successful approaches to the board of RSPacS at times that were likely to be successful.

When I departed from Canberra for London in late July 1982 I did so with some sadness, but I must confess also with a sense of excitement at the prospects ahead of me on the international scene. SDSC had reached a level of maturity at which it could not be stopped or stunted. I had done my bit there for the past 12 years.

It was now time for others to take it in the new directions that followed and to build its strength into what it is today, with four professors to lead it, an assured funding base and a sufficiently strong national and international reputation to compete with the best for financial support and the laurels of academic debate.

5

Reflections on the SDSC's Middle Decades

Desmond Ball

This essay was previously published in the 40th anniversary edition. It is reprinted here in its near original format.

I joined the Strategic and Defence Studies Centre (SDSC) as a research fellow in July 1974. I have remained with it for 32 years, nearly a third of a century, and more than three-quarters of its existence. I became a fellow in October 1980, the first tenured position in the Centre and, at the same time, was officially made the deputy head. I served as head from March 1984 to July 1991. It has been home for almost my entire academic career — such a part of my adult life that it is difficult for me to reflect about it objectively.

The Centre was established at The Australian National University (ANU) by T.B. Millar, a former Australian Army officer, in 1966, when he was a senior fellow in the Department of International Relations, to 'advance the study of Australian, regional, and global strategic and defence issues'. Its initial funding was by way of a grant from the Ford Foundation and, organisationally, it was an independent offshoot of International Relations. For two decades it was the only academic centre concerned with strategic and defence studies in Australia.

Several others were established in the late 1980s and the 1990s, but the SDSC has remained pre-eminent in terms of international reputation and research productivity.

SDSC was headed from 1971 by Bob O'Neill, a former army officer and a senior fellow in International Relations. He presided over the Centre's expansion and rise to international recognition throughout the 1970s and early 1980s. He was already the author of three books: *The German Army and the Nazi Party* (1966), the classic text on civil–military relations in Nazi Germany; *Vietnam Task* (1968), based on his experiences in Vietnam, where he had served as an infantry captain in 1966–67; and *General Giap: Politician and Strategist* (1969), a biography of the North Vietnamese military leader, the architect of the Viet Minh victory at Dien Bien Phu in 1954 and, two decades later, of the defeat of the United States in Vietnam. He was by 1971 regarded as Australia's leading soldier–historian and one of its best military historians ever. O'Neill's major research project during his 11 years at SDSC was his two-volume, 1,300-page official history of *Australia in the Korean War: 1950–53*. Volume 1, on strategy and diplomacy, was published by the Australian War Memorial and the Australian Government Publishing Service in 1981, and Volume 2 on combat operations was published in 1985. Reviewers said that the twin works 'will always be the indispensable reference' on Australia's role in the Korean War.[1]

By 1974, however, when I joined him, O'Neill's talents were already turning to institution-building and project leadership. His first task was to build a critical mass of research posts in SDSC, based on a core staff of longer-term appointments. He promoted the Centre through regular public conferences and by developing contacts with the media. The conferences were usually products of extensive research projects, and usually addressed subjects for the first time in Australia.

Through the mid-1970s, O'Neill obtained financial support for several core posts. In 1973, when Lance Barnard was Minister for Defence in the Labor Government under Gough Whitlam, he secured funding from the Department of Defence for two academic posts; these appointments were for two years, with possible extension to a maximum of five years.

---

1 Desmond Ball, 'Robert O'Neill: A Strategic Career', *Australian Journal of International Affairs*, Vol. 60, No. 1, Mar. 2006, pp. 7–11.

He was then able to move SDSC into the staffing and budgeting system of ANU and obtained an ANU-funded post in 1976 and another two over the next five years. He also forged a strong relationship with the Ford Foundation and, later, the MacArthur Foundation.

I received one of the first two Defence-funded posts, beginning a relationship with Defence that we both often found uncomfortable over the ensuing years. The other post, for work on regional security issues, went to Peter Hastings, the pungent, waggish and quarrelsome journalist, who worked on political and security issues concerning Indonesia and Papua New Guinea. He enjoyed regular access to the office of the then director of the Joint Intelligence Organisation (JIO), as well as conviviality and good wine. He married Jolika Tie, who was our research assistant, in 1981. Two other key members of the Centre at this time, when a critical mass was being put together, were Jol Langtry and Billie Dalrymple. Langtry, another lover of good wine, was the Centre's executive officer from August 1976 to December 1988. A former Army officer who had worked in JIO and army combat development areas, his ability to think of novel strategic and operational concepts was inspirational. Dalrymple was the Centre's secretary from 1977 to 1989. As O'Neill said when she retired, Dalrymple was the crux of a hive of activity, working unstintingly 'with her own special flair and style, smoothing down ruffled feathers when others became agitated, cheering those under pressure and dealing with the outside world with charm and panache'.

O'Neill moved to London to head the International Institute for Strategic Studies (IISS) in 1982. He was already recognised internationally for his leadership qualities, adeptness at collegiate and foundation politics and immense personal integrity, as well as his intellectual work. In 1987 he became the Chichele Professor of the History of War at All Souls College at Oxford University, where he stayed until his retirement in 2001.

I saw O'Neill display not only superb diplomatic skills but also immense integrity and a commitment to academic values. Some of the Centre's work was intensely controversial, as befitting path-breaking scholarship on major national and international issues. Some senior Defence and intelligence officials regarded my own work on US installations in Australia, such as Pine Gap, with great suspicion. While I argued that it was necessary in a democracy for the public to know the purposes and implications of these facilities, a proposition now taken for granted, the senior External Affairs and Defence public servant Sir Arthur Tange complained that I was dangerous and irresponsible, opening up matters that demanded absolute secrecy. He was especially upset because my post was then funded by Defence. O'Neill defended the right of academics to pursue unfettered research. Only when I later became head of SDSC and inherited the files of correspondence between O'Neill and Tange did I fully appreciate the extent of his discourse and the solidity of his refusal to countenance any hint of infringement on the principle of academic independence.

Tom Millar's Return

When O'Neill resigned, the Director of the Research School of Pacific Studies (RSPacS) appointed Tom Millar as acting head, effective from 1 August 1982. He was reappointed head from 14 October 1982. He was then a professorial fellow in International Relations and, of course, had been the founding head of the Centre. The relationship between Millar and myself was proper but not warm. We both held the interests of SDSC in the highest regard, but we were of different personalities and some of my social and political values were difficult for him to abide. He objected to my partner Annabel Rossiter and I living together without being married, which excluded any social relationship between us. He was rigid in some of his views.

Millar was soon embroiled in a bitter controversy about the academic merits of peace research and its place in the Research School. He was strongly supportive of the study of arms control, non-proliferation and disarmament and, soon after resuming the headship, he sought funds from both the Department of Foreign Affairs and ANU to support work in this area. In mid-1983, ANU agreed to fund a post in 'Arms Control, Disarmament and Peace Research' in the Centre, to which Andrew Mack was appointed later in the year. By this time, the Labor Government under Bob Hawke was in office. Over the previous couple of years, Mack (then a senior lecturer in international politics at Flinders University and Australia's leading academic peace researcher) and I had separately been working with the Australian Labor Party's Foreign and Defence Committee on the inclusion of a commitment to the establishment of a Peace Research Centre (PRC) in the party's platform, and talking at length with Bill Hayden about its implementation. Together, I think, we persuaded Hayden, when he became Foreign Minister, to place the new Centre in RSPacS, 'separate from but to cooperate closely with' SDSC. Millar was very upset; he believed that peace research could too easily become 'unprofessional', its rigour compromised by activist agendas, and had to be subsumed within and under the broader subject of strategic studies.[2] He resigned as head of SDSC and returned to International Relations on 31 January 1984. He later became Professor of Australian Studies and head of the Sir Robert Menzies Centre for Australian Studies at the Institute of Commonwealth Studies in London (1985–90). After Millar died in London in June 1994, Coral Bell produced a volume of essays in his honour.[3]

Mack stayed in the Centre for nearly two more years, completing a comprehensive review of peace research in the 1980s,[4] before becoming the first head of PRC in late 1985. Mack and I had a superb working relationship for the next half decade. We were members of each other's advisory committees; we organised joint SDSC–PRC conferences and published joint books and articles; we lobbied within ANU and with US foundations on each other's behalf; and we socialised together a lot, mixing business with pleasure. We were very good friends. Our relationship quickly deteriorated, however, after Mack moved to the chair of International Relations in 1991, an appointment in which I had played a substantial role. We argued vehemently about the relative resources that SDSC and International Relations received from the Research School; Mack now thought that SDSC, although still much smaller than his department in terms of academic staff, was being favoured in Research School budgetary processes. Yet I believe Mack was also jealous of the relative productivity of the Centre and the international reputation that we enjoyed. I saw an unpleasant side of Mack that I had not hitherto appreciated.

---

2   T.B. Millar, *The Resolution of Conflict and the Study of Peace*, Working Paper No. 73, (Canberra: Strategic and Defence Studies Centre, ANU, Jul. 1983).
3   Coral Bell (ed.), *Nation, Region and Context: Studies in Peace and War in Honour of Professor T.B. Millar*, Canberra Papers on Strategy and Defence No. 112 (Canberra: Strategic and Defence Studies Centre, ANU, 1995).
4   Andrew Mack, *Peace Research in the 1980s* (Canberra: Strategic and Defence Studies Centre, ANU, 1985).

Structuring SDSC's Research Agenda

By the time O'Neill left in 1982, the enhanced SDSC had members working on global strategic issues, security developments in the Asia-Pacific region and Australian defence matters, funded through various arrangements, including several visiting fellowships. For planning purposes, I adopted this three-tier structure as a basis for the Centre's further expansion throughout the 1980s, obtaining a core establishment of tenured or 3–5 year appointments to lead the research in the three areas, with an increased number of visiting fellows of various sorts. Conferences remained a major feature of the research programs.

The largest proportion of the Centre's work in the second half of the 1970s and in the 1980s concerned the defence of Australia. SDSC was at the forefront of the conceptual revolution in Australian defence policy from 'dependence on great and powerful friends' to 'greater self-reliance' and from 'forward defence' to 'defence of Australia' that occurred during this period. The first major step in this process was a conference that O'Neill organised on 'The Defence of Australia: Fundamental New Aspects' in October 1976, which was designed to assist policymakers struggling with the transformation of Australia's defence posture. It included papers by leading overseas experts on the concept of 'total defence' and on the strategic and tactical implications of new conventional weapons technologies; by Bill Morrison, who succeeded Lance Barnard as Defence Minister in 1974 and was a visiting fellow in SDSC in 1976–77, on the role of the minister in policymaking since the reorganisation of Defence in 1973–75; on force structure and equipment acquisition matters; and O'Neill's own paper on the development of operational doctrine for the Australian Defence Force (ADF).

SDSC contributed to the development of new ideas concerning many aspects of the defence of Australia, including the command and control of the ADF, the establishment of 'functional' command arrangements for joint operations; reorganisation of the Defence portfolio, such as the establishment of the Defence Council, which was recommended by Millar; greater utilisation of the civilian infrastructure, especially in defence of Australia contingencies; greater appreciation of the challenges of lower level contingencies in northern Australia; regular officer education and training; and particular force-structure issues.

Members of the Centre were credited with an influential role in the government's decision in 1981 to acquire the F/A–18 as the RAAF's tactical fighter aircraft. Costing $4 billion, this was the largest capital program in Australia's history, and turned out to have been the right choice. The core people involved in this work on Australian defence were O'Neill, Langtry and myself, together with Ross Babbage, initially as a doctoral student in International Relations in the mid-1970s and later (1986–90) as senior research fellow and deputy head of SDSC, but we relied greatly on a stream of Defence-funded visiting fellows, including mid-career ADF officers, for their operational and planning expertise.

Babbage was the conceptual leader. He introduced me to thinking about the defence of Australia at Sydney University in 1972–74, where I was a lecturer before joining SDSC, when he did his Master's thesis on 'A Strategy for the Continental Defence of Australia'. I was his supervisor, but he was far more knowledgeable about the subject. I encouraged him to move to Canberra in 1974 and I assisted O'Neill in supervising Babbage when he was a doctoral student. His PhD, published in 1980 as *Rethinking Australia's Defence*, was the seminal study of the subject. His major work when he was in the Centre between 1986 and 1990 was *A Coast Too Long: Defending Australia Beyond the 1990s*, published in 1990.

Many of the ideas developed in the Centre during the second half of the 1970s and the first half of the 1980s, especially those relating to northern defence, were incorporated in Paul Dibb's *Review of Australia's Defence Capabilities* produced for Minister for Defence Kim Beazley in 1985–86, and described by Beazley as 'the most important appraisal of Australia's defence capabilities since the end of World War Two'. Dibb joined International Relations as a senior research fellow in 1981 and the SDSC in 1984, and took leave from SDSC to produce the review.

The field trips we made around northern Australia during the 1980s, using Coastwatch or RAAF aircraft, four-wheel-drive vehicles and river barges, mapping the local civil infrastructure and vital national installations, proffering novel operational concepts for northern defence, and seeing these being tested in large-scale defence exercises, were exhilarating affairs. In addition to my first trip across northern Australia, from Cape York to the Kimberley region of Western Australia, in July–August 1983, I still have vivid memories of trips through the Northern Territory and the Kimberley with Langtry in July–August 1984 and September–October 1985, with Langtry and Babbage up the Tanami Track and through the East Kimberley in October 1986, and around the Torres Strait with Babbage, Langtry and Cathy Downes in May–June 1987. My daughter Katherine, born in 1984, was named in part after the township 320 kilometres south of Darwin, which we had identified as the focal point for the defence of the Top End, and where the first squadron of the new F/A–18 fighters would soon be based. One of the particular northern infrastructure projects for which we became leading proponents was construction of an Alice Springs to Darwin railway connection, and it was pleasing to be invited to Darwin in October 2003 to see the first train come up the line.

The second significant area of work in the Centre, which brought us to international attention, concerned the strategic nuclear balance between the United States and the Soviet Union, and related issues of nuclear proliferation. The first SDSC conference that O'Neill organised, in July 1974, addressed US, Soviet and Chinese strategic nuclear policies and capabilities, nuclear arms control and non-proliferation; it was the first serious examination of the subject in this country. The papers were edited by O'Neill and published by SDSC in August 1975.[5] A 'follow-up' conference on 'The Strategic Nuclear Balance 1975' was held in June 1975, with the edited papers being published by SDSC in May 1976.[6]

My work focused on the operational aspects of strategic nuclear targeting and the controllability of nuclear war, and showed that the mechanisms needed for controlling a nuclear exchange degraded rapidly after only several tens of detonations or a day or so of operations, leading inexorably to full-scale nuclear war. These were heady days, involving frequent sojourns to underground missile silos, the warning centre under Cheyenne Mountain near Colorado Springs, the Pentagon, the US intelligence agencies and the White House. I sat only feet away from the 1.2-megaton nuclear warheads atop the Minuteman ICBMs at Whiteman Air Force Base, each about a hundred times more powerful

---

5   Robert O'Neill (ed.), *The Strategic Nuclear Balance: An Australian Perspective* (Canberra: Strategic and Defence Studies Centre, ANU, 1975).

6   H.G. Gelber (ed.), *The Strategic Nuclear Balance 1975* (Canberra: Strategic and Defence Studies Centre, ANU, 1976).

than the bomb that destroyed Hiroshima in 1945. I also spent a few days at McConnell Air Force Base near Wichita in Kansas, which had 18 Titan II ICBMs, each with 9-megaton warheads, or nearly a thousand times larger than the Hiroshima bomb. In 1982 I worked with a group of recent and current officials from the Central Intelligence Agency (CIA), the National Security Agency (NSA), including two former NSA directors, and the Pentagon, on the vulnerabilities of US nuclear command and control systems. In 1985–86 I was privileged to be a member of a group sponsored by the American Academy of Arts and Sciences to study 'the role of crisis as a precursor to nuclear war and the extent to which the superpowers' command organisations could maintain control over such a chain of events'. The other participants included former president Jimmy Carter, who I had only met once before; former secretary of defense Robert McNamara, who I already knew well; Hans Bethe, the nuclear physicist and Nobel Laureate, and Richard Garwin, the two most brilliant men I have ever met; McGeorge Bundy and Brent Scowcroft, former national security advisors to presidents Kennedy and Nixon, both of whom I also knew well; recent commanders-in-chief; and Condoleezza Rice, a specialist on the Soviet High Command who was working at the Joint Chiefs of Staff.[7] In 1988 I did a report for the US Air Force Intelligence Agency on Soviet signals intelligence (SIGINT) capabilities, the principal source of 'strategic warning' of a nuclear attack.[8] I was in West Berlin on 9 November 1989, when the Berlin Wall was demolished, watching the panicked Soviet intelligence officers based in the Soviet consulate desperately reacting to the loss of some of their covert technical equipment. This work was not only extremely exciting, it also enhanced the profile of Australian strategic studies in the United States.

The third broad area of Centre research in the late 1970s and the 1980s concerned security issues in the Asia-Pacific region. We had a succession of two- to three-year appointments on various aspects of regional security, funded variously by the Department of Defence, ANU and the Ford Foundation. They included Lee Ngok, Donald McMillan, Gary Klintworth and Denny Roy who worked on China; Paul Keal and Peter Polomka on Japan; Greg Fry and David Hegarty

---

7   Kurt Gottfried and Bruce G. Blair (eds), *Crisis Stability and Nuclear War* (Oxford University Press, 1988).
8   Desmond Ball, *Soviet Signals Intelligence (SIGINT)*, Canberra Papers on Strategy and Defence No. 47 (Canberra: Strategic and Defence Studies Centre, ANU, 1989).

on the south-west Pacific; R. Subramanian, S.D. Muni, Sreedhara Rao, Pervaiz Iqbal Cheema and Sandy Gordon on South Asia; and Alan Dupont on Indonesia. Their names are associated with standard reference works in their respective areas.

I became head of SDSC on 7 March 1984, during a period when I was spending lengthy periods overseas — at the Centre for International Affairs at Harvard University, the RAND Corporation in Los Angeles and IISS in London, as well as various places in Washington D.C. — and was contemplating moving to the United States. In March 1987 I was awarded a personal chair, one of six 'special professorships' created in the Institute of Advanced Studies 'in recognition of a high international reputation for distinguished academic work'. I had really wanted to stay at ANU, both because I much preferred living in Canberra to any major city in the United States, particularly now that Annabel and I were having children, and because of the opportunity to devote a lifetime to academic research in the Research School that the personal chair offered.

The working environment in the 1970s and 1980s was more relaxed and sociable than in later years. There was more time for informal discourse between colleagues from different parts of the Research School and, indeed, ANU, perhaps lubricated by good wine on the lawns of the old Staff Centre (Old Canberra House). The contemporary research projects and publications tended, as a result, to be broader and more multi-disciplinary. Books published by Centre members in the 1980s included chapters by Rhys Jones in pre-history, John Chappell in biogeography, Andy Mack and Trevor Findlay in PRC, Hal Hill in economics, Richard Higgott in International Relations, and Jamie Mackie and Ron May in the Department of Political and Social Change. The discussions with Jones led to one of my favourite edited books, *Aborigines in the Defence of Australia*, in which he and Betty Meehan wrote a chapter on 'The Arnhem Salient'.

By the end of the 1980s, SDSC was being consistently ranked among the top 15 or 20 strategic studies centres in the world. In 1990, the review of the Institute of Advanced Studies, chaired by Sir Ninian Stephen, cited SDSC as an illustration of 'how well parts of the Institute's research have met the goals of those who created the ANU'. The Vice-Chancellor, Professor Laurie Nichol, said it 'is one of this University's major success stories'. Bill Hayden, then

Governor-General, said at the SDSC's 25th anniversary conference in 1991 that its influence extended 'well beyond academic cloisters' and that 'this kind of interaction between scholars, policymakers and the broader community was in fact the inspiration behind the establishment of the Institute of Advanced Studies in 1946'.[9]

Defence Funding

The rise to prominence of the Centre during the 1970s and 1980s would not have been possible without the largesse of the Department of Defence. The two posts that O'Neill arranged with Barnard and Tange in 1973–74 were the cornerstones in his building of SDSC. O'Neill also arranged, in 1977, for Defence visiting fellows to come to the Centre for 12-month periods; the first of these, in 1977–78, was Lieutenant Colonel Steve Gower, who worked on 'options for the development of a defence technological strategy' for Australia.[10] Inevitably, however, the dependence on Defence funding, as with all external funding, created difficulties. There were pressures to change some of our research directions and constraints imposed on some of our research activities.

I initially wanted to research decision-making with respect to US facilities in Australia, as part of a larger project on the politics of Australian defence decision-making. I did not have to deal directly with Tange's rage at this notion; that was for O'Neill. But Tange's message that Defence funds should not be used on this subject was clear. When I published *A Suitable Piece of Real Estate: American Installations in Australia* in 1980, I specifically wrote that 'this book was written at home rather than in my office at the Australian National University',[11] in a lame attempt to distance SDSC from it. I was gratified when O'Neill commented at the launching at the National Press Club that he regarded it as an important SDSC product.

---

9   Desmond Ball & David Horner (eds), *Strategic Studies in a Changing World: Global, Regional and Australian Perspectives*, Canberra Papers on Strategy and Defence No. 89 (Canberra: Strategic and Defence Studies Centre, ANU, 1992), pp. xxvi–xxxi, 10–20.
10   S.N. Gower, *Options for an Australian Defence Technological Strategy* (Canberra: Strategic and Defence Studies Centre, ANU, 1982).
11   Desmond Ball, *A Suitable Piece of Real Estate: American Installations in Australia* (Sydney: Hale & Iremonger, 1980), p. 13.

When my own two-year Defence-funded appointment was up in July 1976, Tange refused to renew the post. I am sure that his anger at my appointment was behind his decision. I was fortunately appointed to the SDSC's first ANU-funded post (with O'Neill being officially on the books in International Relations), which O'Neill had secured earlier that year. The other Defence-funded post was never threatened: Peter Hastings was extended until 1977; he was succeeded by Philip Towle in 1978–80, Paul Keal in 1981–83 and Greg Fry in 1983–86.

My most difficult times were with Bill Pritchett, Tange's successor in Defence. He took great umbrage at the work that Langtry and I were doing on Defence's mobilisation planning, where we were finding grave deficiencies.[12] On one occasion he called me over to his office, after some embarrassing revelation by Langtry and I about the inadequate planning machinery, and demanded that I sign a retraction he had prepared. In 1982 he intervened with the Pentagon and then the State Department to persuade the RAND Corporation in Santa Monica (where I was working from time to time as a consultant on US strategic nuclear targeting policy) to curb my access to areas holding the most highly classified materials, effectively making it impossible for me to work in the main building. He argued that my position at RAND, funded by the United States Air Force (USAF), gave credibility to criticisms I was making of the US facilities in Australia, including the USAF's own satellite ground station at Nurrungar in South Australia. I have to say that both RAND and the USAF were supportive throughout this affair. They both initially resisted the pressures from Canberra and the State Department, and both RAND management and successive chiefs of the USAF ensured that our working relationship was maintained for another decade.

Defence support was substantially revamped after Beazley became Minister for Defence. He agreed in May 1985 to new arrangements that were formalised in a memorandum of understanding (MoU) signed by Sir William Cole, secretary of Defence, and Peter Karmel, ANU Vice-Chancellor, in November 1985. Under these new arrangements, Defence funded two three-year research fellow/senior research fellow posts, one in the area of Australian defence and one in the area of regional security, and two three-month visiting fellowship posts

---

12   Desmond Ball & J.O. Langtry (eds), *Problems of Mobilisation in Defence of Australia* (Canberra: Phoenix Defence Publications, 1980).

per year. In early 1987 Defence also agreed to fund two scholarships per year for Australian scholars to undertake the new Master's degree in the Centre. The first appointment in the regional security area was David Hegarty, who joined SDSC in September 1986; the first in the Australian defence area was Cathy Downes, who joined SDSC in January 1987 and who worked on manpower recruitment and development policies and practices for the ADF.

SDSC had 11 Defence-funded visiting fellows from 1986 until 1991, including Michael McKinley, Carl Thayer, Gary Brown, Sandy Gordon, David Jenkins, Alan Dupont and Mohan Malik. A full list, together with their research projects, is in the SDSC December 1991 Newsletter.

In addition, arrangements were instituted for the three Services to send officers to SDSC for 12- to 18-month visiting fellowships. This program was initiated with the Royal Australian Navy (RAN) in July 1984; the first RAN fellow was Commander Simon Harrington, followed by Commander G.F. Smart and then Commander Bill Dovers. MoUs were signed with the Army and the Royal Australian Air Force (RAAF) in January 1990. The first Army fellow was Brigadier Paul O'Sullivan and the first RAAF fellow was Wing Commander Gary Waters. There were 17 Defence and Service fellows at the Centre from 1976 until 1991; they are also listed in the December 1991 Newsletter.

From time to time, Defence also funded visiting fellows on an individual basis. These included Lieutenant Colonel Jim Sanday (1987–88), who had been chief of staff and deputy commander of the Royal Fijian Military Forces before the coup in Fiji in May 1987; Balthasur Tas Maketu (1988–89), who had been secretary of the PNG Department of Defence since 1974; and Denis McLean (1989–90), who had been secretary of New Zealand's Ministry of Defence for the previous decade. We also had several secondments from Defence, including Fred Bennett in 1988–89, who, prior to joining SDSC, had been chief of capital procurement in the Defence, and Barry Roberts, who was seconded to SDSC in November 1987 to provide the quantitative skills required for our new graduate program.

Whereas O'Neill resisted the involvement of Defence in the appointment processes for the Defence-funded posts, I initiated some measures whereby the Defence interest would be heard without compromising academic criteria. I discussed the proposed research topics of applicants for the Defence-funded posts with the appropriate Defence officials, including Pritchett on at least one occasion, but I was not beholden to their preferences and I did not divulge the names of the applicants. In 1990, when Gordon was appointed to the regional security post and Stewart Woodman to the Australian defence post, we even invited Defence to nominate an official to serve as a member of the appointment committees. The appointment of Bob Mathams to the Centre's Advisory Committee in 1985, as noted below, helped improve our credibility in important parts of the Defence establishment.

The Advisory Committee

From the beginning, SDSC had an ardent and assiduous Advisory Committee, chaired by the Director of RSPacS. Its role was 'to advise the Director of RSPacS and, through him, the ANU Vice-Chancellor on matters of policy relating to SDSC; and to advise the Head of SDSC on the Centre's research program'. Its first chair was Sir John Crawford, who played a leading role in the Centre's establishment. Its members were senior academics from elsewhere in ANU, mostly from relevant departments in the H.C. Coombs Building. It had proven helpful to O'Neill as he moved SDSC into the Research School structure, and was later invaluable to me. It supported me in the intra-School politics as we claimed an increasing number of RSPacS posts, reaching four by the late 1980s; and it gave me considerable protection against external pressures, including pressures from Defence. The Director of RSPacS and chair of the Advisory Committee from 1980 to 1993 was R. Gerard Ward; he was supportive of SDSC, for which I will always be grateful, although our relationship began to fray in the early 1990s as budgetary cuts hit RSPacS and he was unavoidably drawn into my conflict with Mack.

Two of the initial members of the committee, Hedley Bull and Bruce Miller, the joint heads of International Relations from 1967 until Bull went to Oxford University in 1977, played important roles in the foundation and early development of SDSC. Miller, together with Crawford, conceived the idea of a Strategic and Defence Studies Centre; he remained on the committee until his retirement in 1987. Bull brought his intellect and the reputation that followed publication of *The Control of the Arms Race*,[13] but he could sometimes be difficult, puffing on his pipe between acerbic comments. At the SDSC conference on Australian defence policy in October 1976, he intervened in a heated discussion about alternative defence-planning methodologies to opine that the whole subject was a waste of time; there were more momentous issues in the world warranting academic inquiry than defending Australia. He did not really believe this, and in fact wrote several articles about Australian defence, but he enjoyed sniping.

Other members of the Advisory Committee in the late 1970s and the 1980s included Harry Rigby, a senior fellow in the Research School of Social Sciences and an internationally recognised scholar on the Soviet Union and especially the Communist Party of the Soviet Union; Jim Richardson, Coral Bell and Geoffrey Jukes from International Relations; Professor Max Corden from the Department of Economics; and Professor Jamie Mackie from Political and Social Change. Mack joined the Advisory Committee when he became head of PRC in 1985. Hugh Smith, from the Department of Government at the Australian Defence Force Academy (ADFA) in Canberra and an expert on civil–military relations, was also appointed in 1985.

I also arranged for Mathams to be appointed to the Advisory Committee. He had headed the Scientific Intelligence Group in the Joint Intelligence Bureau from 1958 until the creation of JIO in 1969, when he became the Director of Scientific and Technical Intelligence in JIO.[14] He played a central role in the establishment of the ground station for the CIA's geostationary SIGINT satellites at Pine Gap. This program gave Australia a central role in maintenance of the global strategic balance, and, at a personal level, forged connections between Australian defence and intelligence officials and the hierarchy of the CIA's Deputy Director, Science and Technology in Langley, Virginia. He was a good friend of SDSC. I first met him when he attended SDSC conferences on the strategic nuclear balance in 1974 and 1975, and came to know him better in the early 1980s. Although he retired

---

13 Hedley Bull, *The Control of the Arms Race: Disarmament and Arms Control in the Missile Age* (New York: Frederick A. Praeger, 1961).

14 R.H. Mathams, *Sub Rosa: Memoirs of an Australian Intelligence Analyst* (Sydney: George Allen & Unwin, 1982); and R.H. Mathams, *The Intelligence Analyst's Notebook*, Working Paper No. 151 (Canberra, Strategic and Defence Studies Centre, ANU, 1988).

in 1979, he was still, in 1985, highly regarded in Defence and the Defence intelligence agencies. He left Canberra for more northern climes at the end of 1989.

Graduate Students

One of the original distinctions between centres and departments in the research schools, along with the injunction against tenured posts, was that only departments were allowed to have doctoral students. However, both O'Neill and I recognised that viable institutions require continuous regeneration. We both supervised doctoral students working on strategic or defence topics in International Relations, including David Horner, Babbage and Ron Huisken, who later took up senior positions in SDSC. Tim Huxley, one of O'Neill's students at the beginning of the 1980s, is now the senior fellow in charge of Asia-Pacific security matters at IISS.

In 1983 RSPacS changed its policy and allowed SDSC to enrol a small number of doctoral students. The first was Andrew Butfoy, who received a Master of Arts in war studies at King's College, London. He joined SDSC in March 1984 and wrote his dissertation on 'Strategic Linkage and the Western Alliance: Nuclear War Planning and Conventional Military Forces'. Butfoy later was a senior lecturer in international relations specialising in security studies at Monash University. Our second was Robert Glasser, who joined SDSC in January 1986 and wrote his dissertation on 'Nuclear Pre-emption and Crisis Stability'. He is now Chief Executive of CARE Australia, after earlier careers at the Los Alamos National Laboratory in New Mexico and AusAID in Canberra. Our third was Matthew Allen, who began his dissertation on 'Processes of Change and Innovation: A Study of the Development of Military Helicopter Doctrines' in February 1987. The fourth and fifth, who both began in 1992, were Leonard Sebastian, on 'Indonesian National Security and Defence Planning', and Nicola Baker, on 'Defence Decision-making Processes in Indonesia, Malaysia and Singapore'. It remained a small program, with usually only around three or four students at any one time.

Members of SDSC had contributed a course on 'Strategic and Defence Studies' to the Master's program in the Department of International Relations since its inception in 1975 and, in 1986, we decided to establish our own graduate program. It was initiated by Babbage, who took overall responsibility for the program in 1987, its first year. Substantial funding for the program was provided by the MacArthur Foundation, including funds for a program coordinator, an administrator, and some scholarships for Asian students. The founding coordinator was Leszek Buszynski, an analyst of Soviet activities in Asia, who joined SDSC in October 1987. He was later assisted by Stewart Woodman, who joined SDSC in December 1990. As well as the two Defence-funded scholarships for Australian students agreed in early 1987, the New Zealand Ministry of Defence also agreed to fund two 'Freyberg Scholarships' for New Zealand students. In November 1989, British Aerospace Australia began funding an annual scholarship. The program was administered by Tina Lynham, whose devotion to the students was wonderful.

Officer Education and Development

Beginning under O'Neill's tenure, members of SDSC have played important roles in the evolution of ADF officer education and development and, more directly, in the formulation and presentation of the strategic studies components of the courses at the major Service training institutions, in that time the Joint Services Staff College (JSSC) and the Army, RAAF and RAN staff colleges. In 1977, for example, SDSC was requested by the Army's Regular Officer Development Committee (RODC) to prepare a paper on 'the future operational requirement and officer development', which influenced the final report.[15]

SDSC members were active proponents of a single, integrated institution for officer cadet training, of the sort embodied at ADFA. I was a member of the ADFA Council, appointed by the Minister for Defence, from 1985 to 1991, when I was succeeded by Paul Dibb.

Members of SDSC had an especially close affiliation with the JSSC at Weston Creek. O'Neill was a member of the Dunbar Committee, chaired by Deputy Vice-Chancellor of the ANU, which reorganised

---

15   Desmond Ball, Ross Babbage, J.O. Langtry & R.J. O'Neill, *The Development of Australian Army Officers for the 1980s*, Canberra Papers on Strategy and Defence No. 17 (Canberra: Strategic and Defence Studies Centre, ANU, 1978); and Desmond Ball, 'The Role of the Strategic and Defence Studies Centre', in Hugh Smith (ed.), *Australians on Peace and War* (Canberra: Australian Defence Studies Centre, Australian Defence Force Academy, 1987), pp. 77–81.

the JSSC's curriculum and teaching methods in 1975–76. I succeeded O'Neill as Academic Adviser to JSSC after his departure in 1982. I was a member of its Board of Studies from 1986 to 1991, when I was succeeded by Horner who had been on the Directing Staff at JSSC from 1988 to 1990. All of the Centre's academic staff and many of its visiting fellows lectured at JSSC.

Members of SDSC also lectured and assisted with curriculum reform at, and served on the boards of, the Service staff colleges. Horner attended the Army Command and Staff College at Queenscliff in 1983, and was a regular lecturer there from 1983 to 1992. Babbage and I assisted the RAAF Staff College with the major review of its syllabus in 1989–90, after which I served for five years on its Board of Studies. Horner also served on the Australian Capital Territory (ACT) Accreditation Board in connection with graduate accreditation of courses at both the Service colleges and JSSC.

Members of SDSC were extensively involved with the Australian College of Defence and Strategic Studies (ACDSS), which provided education and training for senior ADF officers and civilian officials from 1995 to 2000. I was a consultant to the Chiefs of Staff Committee when ACDSS was proposed and, in 1993–94, several SDSC members were in continuous dialogue with the founding Principal, Air Marshal Ray Funnell, and the Directing Staff. Dibb invariably delivered the 'opening address' to each year's intake, and he and Woodman were responsible for the module on defence decision-making and the policy advisory process.

By the early 1990s, several SDSC members began advocating rationalisation of the staff college system and co-location of the Service staff colleges into a single complex together with the JSSC at Weston Creek. In February 1995, Dibb, myself, Horner and Woodman testified 'in camera' to the Joint Standing Committee on Foreign Affairs, Defence and Trade, which was inquiring into the provision of academic studies and professional military education to ADF officers. We reckoned that between the four of us we had something like 90 years' experience in officer education or professional matters within Defence. We argued, with some slight differences of 'nuance and opinion', that rationalisation and co-location of the various colleges was essential to enable mid-level officers to think about operational and strategic matters in joint terms, that there was much duplication at the separate colleges with respect to both facilities and curricula, and that a single Australian Defence College (ADC) would be more cost-effective. In 1997–98, Dibb was commissioned by Defence to review the higher defence education requirements of the new ADC, and to submit proposals for its educational objectives and curriculum.

ASIO and the KGB

Some of our work was regarded with deep suspicion by a number of senior Defence and intelligence officials who believed that defence policy and national security should be a secret domain. My own work on US installations in Australia, especially the Pine Gap station, caused the greatest anxiety.

The Australian Security and Intelligence Organisation (ASIO) began watching me soon after I joined SDSC. It started a file on me in April 1966, when I was a second-year undergraduate at ANU and was involved in an anti-Vietnam War demonstration, and thereafter had regularly reported my participation in student protest activities. In 1969, when I was a doctoral student in International Relations, ASIO became concerned about the interest I was taking in the establishment of the CIA's satellite control station at Pine Gap. In December 1968, Robert Cooksey, a lecturer in the Department of Political Science, published an article on Pine Gap in *The Australian Quarterly*, in which he acknowledged my assistance. In April 1969, at the request of Sir Henry Bland, Secretary of the Department of Defence, Sir Charles Spry, the Director-General of ASIO, prepared a report on Cooksey that speculated about our motivations. Spry asked ASIO's ACT office to 'fully identify' me.[16] I assume that Bland was prompted by Raymond Villemarette, the CIA's Chief of Station in Australia. I had not known that the CIA was the US agency in charge of Pine Gap until it was revealed by Brian Toohey in the *Australian Financial Review* in November 1975, a disclosure described by Tange as 'the gravest risk to the nation's security there has ever been', prompting the 'security crisis' in the week preceding the downfall of the Whitlam Government. The CIA, however, was concerned that my research might reveal both its role and the existence of its geostationary SIGINT satellite program

---

16   National Archives of Australia (NAA), 'Ball, Desmond John: B/78/42, Volume 1', CRS A6119.

(then codenamed Rhyolite). Although I learned about the Rhyolite program in 1977, following the arrest and trials for espionage of Christopher Boyce and Andrew Lee ('the falcon and the snowman'), I was uncertain about whether the Rhyolite program was Pine Gap's only function until Hayden confirmed it for me in April 1981, after he returned from a tour of the facility. Soon after he became Foreign Minister, Hayden publicly affirmed that Pine Gap was a CIA operation.

By February 1975, six months after I had joined SDSC, ASIO had compiled a preliminary list of my 'contacts'; it was noted that 'the list is not comprehensive as there are additional names on which follow-up action is required'.[17] In January and February, 'a usually reliable source' reported to ASIO that one of my 'contacts' whom I was 'cultivating' was Kevin Foley, a former RAAF officer. The 'source' could not 'reconcile Foley's political beliefs with those of Ball'. In fact, Kevin and I were good friends. We had done our PhDs in international relations at the same time, when very few students were working on defence issues, and we shared many social interests. Recently declassified ASIO files suggest that Millar was the source. Other more personal information was reported or usually misreported by Andrew Campbell, who worked for ASIO in Canberra at various times from 1973 to 1979.

It was reported in September 1980 that my office had been searched and bugged, my files and diaries photographed and my telephone tapped by ASIO as part of 'Operation Answer'.[18] As well as in the late 1970s, it probably also happened on other occasions in the 1980s and 1990s. Operation Answer was reportedly mounted under the pretext of 'counterintelligence and counterespionage', but it was really designed to ascertain my 'links with the Canberra press gallery', and in particular whether I was a conduit for leaks of classified Defence documents to Toohey and William Pinwill, both of whom were journalist friends of mine. I was at IISS in London when the story appeared in the *Australian* in September 1980, but O'Neill sent me a telex with the relevant paragraphs excerpted. He said that he was 'treating [the] issue as of [the] utmost gravity for [the] integrity and academic freedom of SDSC', and that he had asked the ANU Vice-Chancellor, Anthony Low, to take the matter up with the head

---

17   NAA, 'Ball, Desmond John — Miscellaneous Papers', CRS A6119.
18   Richard L'Estrange, 'ASIO Agents Spy on Suspect ANU Scholar', *Australian*, 3 Sep. 1980, p. 2.

of ASIO, Justice Woodward, 'to establish truth or falsity'. I do not know what, if anything, eventuated from this. In May 2004, a former senior officer in ASIO's Counterespionage Branch, who had been involved in spying on me, wrote me a long 'confession'. He noted that 'it had been said that you were a dangerous radical, against the Vietnam War, and a drinker in possession of SIGINT material smuggled outside of a controlled area', but that I was eventually 'rightfully cleared'.

News Weekly, produced by the National Civic Council, reported in 1999 that two SDSC 'directors' had been ASIO sources and that, 'for over two decades, the KGB has regarded SDSC as a key target area in which they can recruit agents of influence and access agents' in Defence and the intelligence community.[19] It said that Lev Sergeyevich Koshlyakov, the energetic KGB Resident in Canberra from 1977 to 1984, was 'well known to key senior Centre staff'. Koshlyakov's cover was Press Attaché, and he would often visit the National Press Club, where I used to imbibe in my drinking days. He was lively, amiable, and reputedly very adept, although I never had a serious conversation with him. *News Weekly* said that Koshlyakov 'frequented' ANU and SDSC, but I do not believe he ever visited the Centre. The only time I recall seeing him outside the Press Club was at a rock concert in front of Old Parliament House in, I think, 1983; Annabel and I were sitting on a blanket when Koshlyakov, in his jeans and leather jacket, joined us for a few minutes.

Two other Soviet officials in this period who did visit both International Relations and SDSC, to attend seminars and to talk with staff members, were Igor Saprykin and Yuri Pavlov. They were Foreign Ministry officers and both later served as Soviet/Russian ambassadors, but several members of the Soviet Affairs Group in ASIO's Counterespionage Branch thought that Saprykin at least was a KGB officer. Dibb got to know them fairly well. He was tasked by his friend, Don Marshall, then the head of ASIO's ACT Regional Office, to cultivate them so as to discover their views on issues concerning the central strategic balance and to discern their real interests and priorities, and possibly persuade one or other of them to defect. I do not believe that I met either of them.

---

19 'Spy Scandal Throws Spotlight on Intelligence Community', *News Weekly*, 5 Jun. 1999, p. 8.

Protests

On the other hand, we were also accused by political activists of various sorts of being agents of the 'military-industrial complex'. Demonstrations were held in protest against many of our conferences, sometimes directed at the participation of particular ministers or overseas speakers and sometimes at our subject matter. On two occasions, hundreds of protesters tried to physically break up the proceedings, once in the Coombs Theatre in November 1989 when the subject was 'New Technology: Implications for Regional and Australian Security' and the other in the Law Theatre in November 1991 on Australia and space. They were misplaced affairs, given the broad and fundamental nature of the conference agendas and the reputations of the overseas participants as leading critical thinkers, and really quite insipid compared with protests against the Vietnam War or nuclear weapons that I had been involved in organising.

The 1990s

Dibb succeeded me in July 1991 and became the Centre's longest serving head, passing O'Neill's tenure (1971–82) by a few months. Frustrated with the administration associated with the position — which was probably less arduous than in more recent times — but for which I was clearly unsuited, I was also anxious to spend less time wearing a suit and tie and more time fulfilling the research commission of my personal chair.

I began sounding Dibb out about possible SDSC headship in mid-1989. Somewhat to my surprise, for this involved a major career change and a commitment to academia rather than a passing stay, Dibb warmed to the idea. In July 1989, at the National Defence Seminar at Canungra, which was sponsored by Beazley as Minister for Defence and General Peter Gration as Chief of the Defence Force (CDF), I asked Beazley what he thought about the proposition, and he gave it his blessing.

In addition to authorship of the Dibb Review in 1985–86, Dibb had served as head of the National Assessments Staff in JIO, the forerunner of the Office of National Assessments in 1974–78, Deputy Director (Civilian) of the JIO in 1978–81, Director of JIO in 1986–88, and Deputy Secretary of the Department of Defence responsible for strategic policy and intelligence from 1988 until 1991. He had also had two previous tenures in the H.C. Coombs Building: as a research fellow in the Department of Political Science in the Research School of Social Sciences in 1967–70, and a senior research fellow in International Relations in 1981–84 and then SDSC in 1984–86, where he wrote the prescient study *The Soviet Union: The Incomplete Superpower* and served as deputy head and, oft-times, acting head.

His remarkable background was eloquently described in the references we solicited for his proposed appointment to a special professorship at ANU and headship of SDSC. Tange, commenting on Dibb's 'rare versatility', said that, as Deputy Director (Civilian) and later Director of JIO and Deputy Secretary (Strategy and Intelligence), he had 'done much to redirect the activities of the intelligence community to matters more closely related to the practical defence interests of the country', that on defence policy issues 'there is none inside or outside the Defence Community better equipped at present to understand the issues in contention and the policy choices', and that Dibb had exhibited remarkable 'courage in arguing with the Services about their own business [i.e. weapons acquisition]'. He also, I might add, could not resist using his reference for Dibb to make some caustic remarks about myself, saying that I had evinced 'some imbalance in the choice of subjects of study', including the US facilities in Australia 'which successive American and Australian Governments have deemed it a national interest' to keep secret, and expressing relief that I would no longer be heading SDSC. Gareth Evans said that 'Dr Dibb's intellectual capacities … are among the most outstanding of the public servants I have encountered in this area of Government'. Gration commented on Paul's 'intellectual rigour', noting that he had 'a unique blend of academic experience and real life strategic policy making, where theoretically attractive concepts have to be tempered with practical realities' and that he had 'a mature understanding of the capabilities, aspirations and limitations of the armed forces as instruments of national policy'. Admiral Ron Hays, who had just retired from the post of US Commander in Chief, Pacific, said he was 'by American standards, a national asset'. Michael McGwire of the Brookings Institution in Washington D.C. said that he had 'earned a first class [international] reputation'. By 1991 Dibb had written five books, four major government reports, and some 100 chapters and articles in scholarly books and journals.

Dibb's accession to the headship coincided with the collapse of the Soviet Union and the end of the Cold War. As a result, he had to manage a wholesale transformation in the Centre's research agenda. The post–Cold War issues were more disparate and diffuse. A new core of academic staff was assembled, consisting, in addition to Dibb and myself, of Bell, Horner, Dupont and, since 2001, Huisken and Clive Williams. Bell became a visiting fellow in SDSC in 1990. Truly indefatigable, she was Professor of International Relations at the University of Sussex in 1972–77 and returned to Australia to spend the next 11 years as a senior research fellow in International Relations, pursuing her passion for comprehending and explaining the fundamental power dynamics of the international system. In the decade and a half that Bell spent with SDSC, she has produced more than half a dozen insightful books and monographs, most recently *A World Out of Balance: American Ascendancy and International Politics in the 21st Century* (2003). Horner, a former Army officer with wide command and staff experience, is Australia's leading military historian. He joined the Centre as its Executive Officer in September 1990, transferred to a fellow in 1994, and a Defence-funded post of Professor of Australian Defence History in 1999. Horner won the J.G. Crawford Prize for the best PhD in ANU in 1982. Huisken was a visiting fellow in the Centre in 1976–77, and returned as a senior fellow after more than two decades in the departments of Foreign Affairs and Trade, and Defence, where he was responsible for arms control issues and the Australia–US defence relationship.

About half of the Centre's work became devoted to Asia-Pacific security matters. Dibb produced the classic studies of the balance of power in the Asia-Pacific region and the revolution in military affairs (RMA) in Asia, as well as the US–Australia alliance. We developed many of the original practical proposals for regional security cooperation in the early 1990s, a lot of which were quickly adopted by the new Association of Southeast Asian Nations (ASEAN) Regional Forum (ARF). SDSC was one of the 10 regional strategic studies centres that, in 1992–93, founded the Council for Security Cooperation in the Asia-Pacific (CSCAP), the premier 'second track' organisation in this part of the world, which now has 22 member committees in 22 countries (with the Australian committee served by a secretariat in SDSC), and that, through its steering committee meetings, study groups and general conferences provides an institutionalised mechanism for continuous activity for promoting regional security cooperation.[20]

SDSC members also explicated a broader conception of security to encompass economic, environmental and other so-called 'non-traditional' threats in addition to the traditional military focus. Dupont's path-breaking book, *East Asia Imperilled: Transnational Challenges to Security*, analysed over-population, deforestation and pollution, global warming, unregulated population movements, transnational crime, virulent new strains of infectious diseases and a host of other issues that could potentially destabilise East Asia. There was an increasing appreciation of the importance of 'human security' as opposed to state security, as reflected in some of my own work on security issues in the Thailand–Burma borderlands.

SDSC took some hard knocks in the 1990s, although its international reputation remained undinted. It suffered from the vicissitudes of dependence on external funding from external sources, and especially Defence, which, at its height at the beginning of the decade, amounted to nearly half of the SDSC budget. More painfully felt were cuts in the Centre's ANU funding and a shift in Research School priorities, which decimated much of its work on Australian defence. It was severely damaged by the move off-campus to Acton House in 1992. This occurred partly at our instigation, as we had PhD students and visiting fellows spread around several buildings and were desperate to bring everyone together. In practice, we found sub-standard premises and intellectual isolation. In October 1999 we moved to the Law Building, which at least had the great benefit of bringing us back onto the campus and close to the H.C. Coombs Building. There was a palpable air of exuberance when we returned to Coombs in September 2004. It was a real homecoming. We were excited about the prospect of daily encounters with colleagues who we had too rarely seen; the closer interaction has already brought cooperative research initiatives and joint publications between SDSC staff and other members of the Research School of Pacific and Asian Studies (RSPAS).

---

20 Desmond Ball, *The Council for Security Cooperation in the Asia Pacific: Its Record and its Prospects*, Canberra Papers on Strategy and Defence No. 139 (Canberra: Strategic and Defence Studies Centre, ANU, Oct. 2000).

The return coincided with other major SDSC developments, producing a sense of regeneration. We have accorded a high priority to educating and training a new generation of strategic thinkers, which has involved greatly expanding our PhD program and developing a new Master's program, directed most ably by Robert Ayson, who himself completed his Master of Arts in the Centre in 1988–89. Dibb reached retirement age in October 2004 and became an Emeritus Professor. Hugh White was appointed head in November 2004. He had previously been Deputy Secretary of the Department of Defence (Strategy and Intelligence). He was the primary author of the Government's Defence White Paper published in 2000, and had been the founding Director of the Defence-funded Australian Strategic Policy Institute (ASPI) in 2001–04. He was attracted to SDSC by our international reputation, but also by the intellectual freedom enjoyed in academia and the depth and breadth of expertise about our region that prevails in RSPAS.

Strategic and defence studies are not popular areas of academic activity. To some critics, the study of war is macabre. Some of our former colleagues in the Coombs Building used to refer to members of SDSC as 'bomb-fondlers', not always in jest. Work on defence planning is regarded as antithetical to the universalism of scholarship. Policy-relevant work is regarded by some as serving the interests of defence and foreign affairs bureaucracies and military establishments, and supporting state power more generally. Academic papers by colleagues elsewhere in ANU have referred to us as 'prostitutes'. Some critics have argued that SDSC should be moved from ANU to Defence.[21]

However, we could not do our job in Defence. Compared to the Coombs Building, we could expect more luxurious facilities and fabulous resources; but we are, at heart, 'defence intellectuals'. I would simply find it unbearable to work in Defence or under any direct or indirect official instruction. The majority of my colleagues in SDSC have spent large parts of their careers in the higher echelons of Defence or the intelligence agencies, but they come to SDSC because of the freedom to think and write independently, critically and objectively, untrammelled by prevailing government policies or bureaucratic interests. Strategic and defence issues are among the most vital issues of public policy; defence capabilities are also enormously expensive.

---

21   David Sullivan, 'Professionalism and Australia's Security Intellectuals: Knowledge, Power, Responsibility', *Australian Journal of Political Science*, Vol. 33, No. 3, 1998, pp. 421–40.

They warrant intensive and rigorous scrutiny and informed public debate, at least as much as health, economic, welfare, environmental or other national issues. SDSC remains the leading academic centre in Australia capable of providing this systematic scrutiny and informing debate.

Plate 23 Dr Fedor Mediansky, Professor Paul Dibb and Professor Desmond Ball, at an SDSC conference, c. 1984

Plate 24 Leszek Buszynski and Ross Babbage, senior research fellows, establishing the first SDSC Master's Program, 1987

Plate 25 Tina Lynham, administrator, SDSC Graduate Program, 1987–98

Plate 26 Denis McLean, visiting fellow, and Jena Hamilton, research assistant (1989–2002), 1989

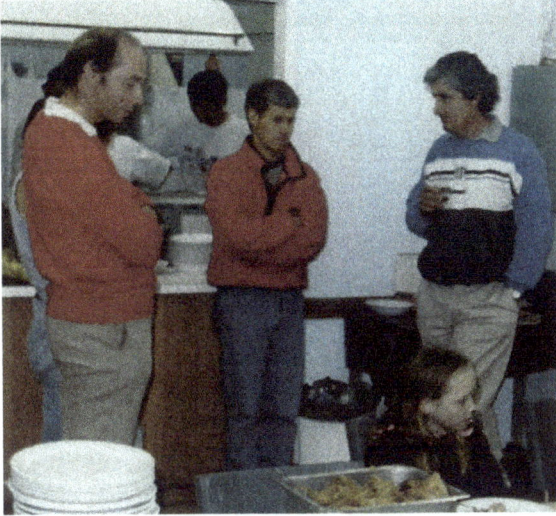

Plate 27 Stewart Woodman, Sandy Gordon and Mike Gilligan,
senior research fellows, Kioloa, 1992

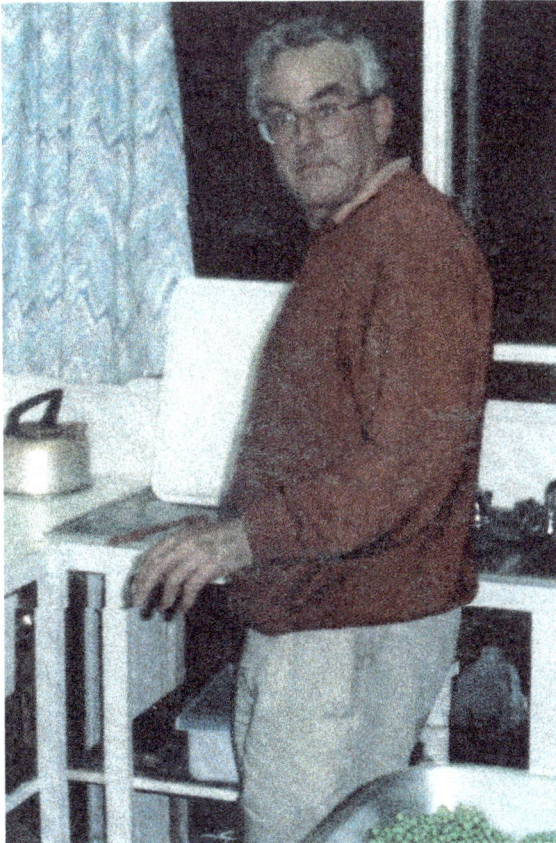

Plate 28 Professor Paul Dibb, head of SDSC, Kioloa, 1992

Plate 29 Elza Sullivan, secretary; Helen Hookey, research assistant; and Liu Jinkun, visiting fellow, Kioloa, 1991

Plate 30 Dr L. Buszynski; Professor R.G. Ward, director of the Research School of Pacific Studies; Mr Kenneth R. Peacock, Rockwell Australia; and Professor Desmond Ball, October 1992

Plate 31 Alexander Downer, Minister for Foreign Affairs and Trade;
Professor Paul Dibb and Professor Robert O'Neill, at SDSC's
30th anniversary conference, 1996

Plate 32 Professor Robert O'Neill and Professor Paul Dibb, Yukio Satoh,
Japanese ambassador to Australia, and Professor Desmond Ball,
at SDSC's 30th anniversary conference, 1996

Plate 33 16th Council for Security Cooperation in the Asia Pacific (CSCAP) Steering Committee meeting, Canberra, December 2001

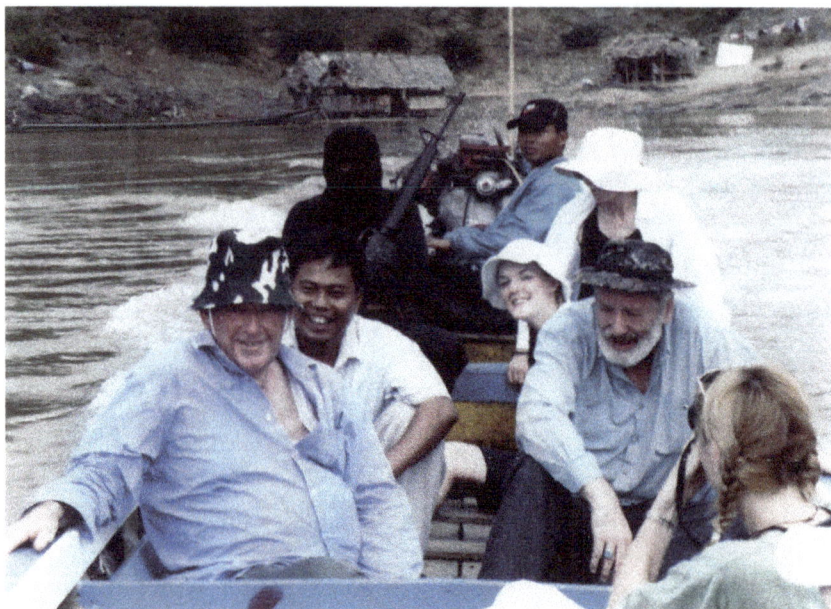
Plate 34 Colin Plowman, Tun Tun Aung, Professor Desmond Ball and a Thai army ranger, on the Salween River, June 2002

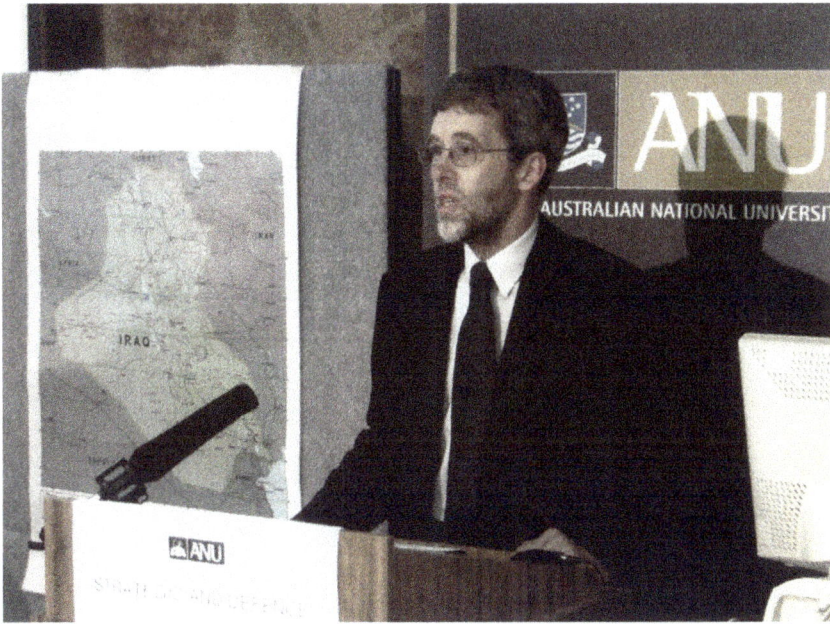

Plate 35 Dr Alan Dupont, senior research fellow, at an SDSC
media briefing on the Iraq War, March 2003

Plate 36 Professor Desmond Ball, Dr Ron Huisken, senior research fellow,
and Ray Funnell, at an SDSC media briefing on the Iraq War, March 2003

Plate 37 Professor Paul Dibb, the Hon. Kim Beazley, and Professor Hugh White, head of SDSC, at Dibb's retirement dinner, May 2005

Plate 38 Professor James Fox, director of the Research School of Pacific and Asian Studies, Professor Paul Dibb and Professor Hugh White, at Dibb's retirement dinner, May 2005

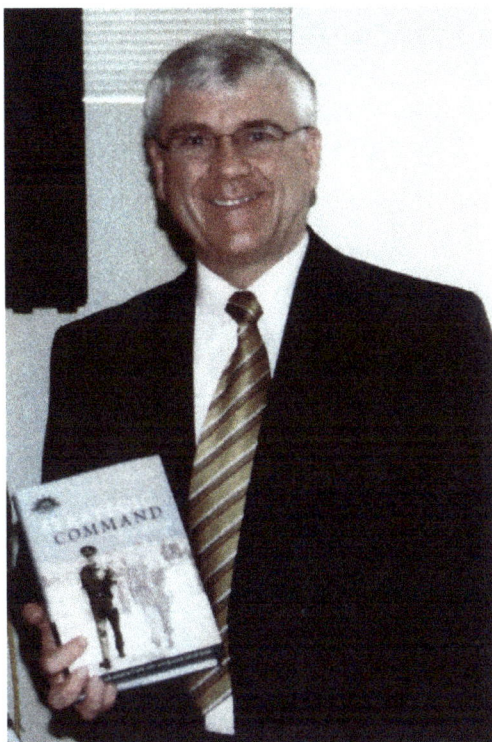

Plate 39 Professor David Horner at the launch of his book
Strategic Command, 6 June 2005

Plate 40 Meredith Thatcher, research assistant; Dr Coral Bell, visiting
fellow; Dr Brendan Taylor, post-doctoral fellow; and Betty McFarlane,
at the investiture of Bell's Order of Australia (AO), Government House,
Canberra, 2 September 2005

Plate 41 The Graduate Studies in Strategy and Defence (GSSD) team:
Dr Chris Chung, deputy director; Ms Ping Yu, program administrator;
Dr Robert Ayson, director; Mrs Sarah Flint, deputy program administrator;
and Dr Brendan Taylor, lecturer, April 2006

Plate 42 SDSC staff with the GSSD students, Kioloa, April 2006

6

SDSC in the Nineties:
A Difficult Transition

Paul Dibb

This essay was previously published in the 40th anniversary edition. It is reprinted here in its near original format.

In the 1990s, the Strategic and Defence Studies Centre (SDSC) underwent a difficult transition. The first 25 years of the Centre's work were concerned with the Cold War and its global and regional impact on Australia. In 1991, however, the Soviet Union disappeared and the Cold War suddenly ended. SDSC had to adjust rapidly to an altered international strategic environment and to new subjects for strategic analysis. In the mid-1990s, the government in Canberra changed and there was a move away from the defence ideas that the Centre had played a key role in developing from the early 1970s. During this decade, The Australian National University (ANU) also experienced significant financial difficulties, which had a serious impact on the SDSC budget and its ability to fund research on important issues.

The terrorist attacks on the United States of 11 September 2001, however, dramatically changed the international security environment and posed new challenges for the Centre's research and teaching agenda. It also brought about new sources of revenue. Thus, the early years of the new century have heralded a much brighter prospect for SDSC and it now finds itself in a much stronger position, both intellectually and financially, than ever before.

This chapter examines how SDSC handled the difficult transition from the end of the Cold War in 1991 to the so-called 'War on Terror' in 2001. It looks at what happened internationally and SDSC's response. It also analyses what happened in Australia and what this meant for the Centre's finances and research priorities. The central idea is to give the reader an understanding of the academic challenges facing SDSC, both in research and postgraduate teaching, and a feel for the problems of managing a small but prestigious institution in a period of difficult transition. This chapter draws heavily on the Centre's annual reports and regular newsletters over the decade of the 1990s, as well as my personal reflections as head of the Centre from 1991 to 2003.

What Happened in the World and SDSC's Response

On Christmas Day 1991, the red hammer and sickle was lowered from the Kremlin and replaced by the white, blue and red flag of Russia. Only with hindsight does the Soviet collapse appear predictable. Even Mikhail Gorbachev did not appear to understand fully what he was doing. And as Stephen Kotkin points out, it was amazing that this hyper-militarised Soviet Union did not attempt to stage even a cynical foreign war to rally support for the communist regime.[1] Even if Soviet leaders had calculated that they were doomed, they could have wreaked terrifying havoc out of spite, or engaged in nuclear blackmail. Historically, such a profoundly submissive capitulation, as took place in the Soviet case, was a rarity. In the past, as Paul Kennedy has observed, none of the overextended, multinational empires ever retreated to their own ethnic base until they had been defeated in a great power war.[2]

---

1   See Stephen Kotkin, *Armageddon Averted: The Soviet Collapse, 1970–2000* (New York: Oxford University Press, 2001).

2   Paul Kennedy, *The Rise and Fall of the Great Powers* (London: Fontana Press, 1989), pp. 567, 694.

I dwell on this point for two reasons. The first is that the international academic community of Sovietologists, including those in Australia, did not predict that the Soviet Union was in deep trouble until the end (the intelligence communities in the United States, the United Kingdom and Australia had the same problem). The explanation, in my view, was over-specialisation — the inability to 'see the wood for the trees'. The second reason is that the Soviet Union was my focus of intelligence and academic study for 20 years, until 1985. It is commonplace for me to hear that the book that I published in 1986, entitled *The Soviet Union: The Incomplete Superpower*, predicted the end of that country. It did not: it merely pointed out that the Soviet Union was a power in difficulty and that, if Gorbachev's reforms failed, the Soviet Union risked falling out of the ranks of the world's great powers.[3]

Throughout much of the Cold War, Professor Des Ball made an outstanding contribution to our understanding of both Soviet and American nuclear targeting doctrine. In the early 1990s, he continued this by publishing articles on Soviet signals intelligence (SIGINT), and on the future of the global strategic balance and current developments in US strategic nuclear targeting policy. Coral Bell continued to work on great power and alliance issues: in 1994 she wrote about the new world order and the Gulf War, edited the book *The United Nations and Crisis Management: Six Studies* and several works on the Cold War in retrospect. The increasing dominance of the United States in world affairs was reflected in a book by Bell on the American ascendancy in international politics and diplomacy in the unipolar period, articles by Ball on the prospects for the US–Australia alliance, and works by myself on the question of whether America's alliances in the Asia-Pacific region would endure.

The biggest change, however, in the SDSC research agenda during the 1990s was a refocusing on the imperatives of understanding the new challenges to Australia's regional security. The end of superpower confrontation meant that much more attention could now be paid to the question of improving regional security dialogue and developing a new security architecture. A new range of issues surfaced in the regional strategic agenda, including major challenges that changes in the nature

---

3   Paul Dibb, *The Soviet Union: The Incomplete Superpower*, 2nd edn (London: Macmillan Press, 1988 [1986]), pp. 267, 278.

of conflict presented to many nations in the development of their armed forces, as well as an increased demand for high-level education on strategic and defence policy issues in an era with no clear threats.

One central development, led by Ball, was the creation of the Council for Security Cooperation in the Asia-Pacific (CSCAP). This remarkable 'track two' initiative was created in 1993 and is now a well-established unofficial contribution to security confidence-building in the region. It should be recorded that Ball was not only instrumental at the creation of CSCAP but that he has laboured mightily for well over a decade now to ensure its success both at the regional level and within Australia.

Ball's publications on this subject began in 1991 with an analysis of confidence- and security-building measures as building blocks for regional security and continued through to the late 1990s with studies on preventive diplomacy and security cooperation in the Asia-Pacific region and studies on the evolving regional security architecture.

About 60 per cent of SDSC in-house publications (Canberra Papers on Strategy and Defence and Working Papers) in the decade of the 1990s were on regional security subjects. When in 1991 SDSC celebrated its 25th anniversary with a conference on 'Strategic Studies in a Changing World', it was significant that there was a particularly strong focus on north-east Asia, South-East Asia and the South Pacific. It was especially notable that all the heads, or deputy heads, of the strategic institutes in the Association of Southeast Asian Nations (ASEAN) region were at the conference and presented papers. It is also worth noting what the Governor-General, Bill Hayden, said about the Centre in opening the conference:

> During some three decades of public life, I know that the Parliament, the military, the public service and the media have all benefited greatly from access to the informal views of the Centre ... one very positive aspect of your influence has been to inject the qualities of intellectual rigour, factual analysis and reasoned argument into a subject all easily prey to prejudice and superstition.[4]

---

4   *Annual Report 1991* (Canberra: Research School of Pacific Studies, ANU), pp. 22–23.

The proceedings of the 25th anniversary conference were edited as a book by Ball and David Horner. In that year, the Centre also initiated new research on India and was fortunate to secure the services of Sandy Gordon who, in 1995, produced the book *India's Rise to Power in the Twentieth Century and Beyond*.

The furtherance of security dialogue in the region and the development of concrete ideas for security-building measures gathered momentum. In 1993 I chaired the first security dialogue between Australia and China on behalf of the Department of Foreign Affairs. In 1994 I also chaired an informal meeting of 18 ASEAN Regional Forum (ARF) countries (again on behalf of Foreign Affairs) on practical measures for military and security cooperation and, in the same year, I published a jointly authored monograph with the Foreign Minister, Senator Gareth Evans: *Australian Paper on Practical Proposals for Security Cooperation in the Asia-Pacific Region*.

A major external review in 1995 of the Research School of Pacific and Asian Studies (as the Research School of Pacific Studies had been renamed the year before) noted that SDSC contributed to the Research School's 'high public visibility and considerable involvement with governments and non-government organisations in the region'.[5] The Centre continued to remain at the forefront of developing new concepts for security planning in the region and, partly due to Ball's hard work, in 1996, ARF accepted that CSCAP had a legitimate role in promoting regional security. That year, SDSC also held its 30th anniversary conference on 'The New Security Agenda in the Asia-Pacific' jointly with the International Institute for Strategic Studies (IISS). The conference was addressed by the Minister for Defence, the Minister for Foreign Affairs, the Shadow Minister for Defence, the Chief of the Defence Force and other senior officials, as well as the Chairman of the Council of IISS and its Director.

In 1996 SDSC was host to the 6th CSCAP Steering Committee in Canberra. The meeting was attended by over 50 representatives and observers from the Asia-Pacific region and Europe. In that year, Ball edited the book *The Transformation of Security in the Asia-Pacific Region*, as well as co-authoring *Presumptive Engagement: Australia's Asia-Pacific Security Policy in the 1990s*. Research on regional security

---

5   *1995 Annual Report* (Canberra: Research School of Pacific and Asian Studies, ANU), p. 79.

was steadily expanded, with assistance from Department of Defence funding, to include the effect of population movements, security problems caused by environmental concerns, the relevance of RMA (revolution in military affairs) to regional countries, regional defence decision-making, the US–Japan relationship, Indonesian defence developments, China's foreign and defence policies, and developments in North Korea. SDSC gave assistance to regional countries with the preparation of defence white papers (for example, those of Thailand, the Philippines, Vanuatu and Papua New Guinea). A number of the Centre's academic staff wrote in journals such as *Survival* about the strategic implications of the Asian economic crisis in 1997 and 1998, while Alan Dupont wrote an Adelphi Paper on the effect of the environment on security in the region.

In 1999 Ball began his fine-grained research into security problems on the Thai–Burma border, which is the definitive work on this subject. The East Timor crisis in that year, and, more generally, the problems in Indonesia, dominated much research, with SDSC staff writing and commenting on the Indonesian armed forces, Indonesian politics, and the effect on security relations with Australia of Australia's intervention in East Timor. In 2000 Ball was honoured by being appointed co-chair of CSCAP as a whole. Particularly important publications in that year were his book *Death in Balibo: Lies in Canberra* and my co-edited book *America's Asian Alliances*, as well as Dupont's major book, *East Asia Imperilled: Transnational Challenge to East Asia's Security*, published the following year.

At the time of writing, in SDSC's 40th year, the Centre has adjusted to the new security challenges of the post–Cold War era and established a reputation for being the leading academic authority in Australia on Asia-Pacific security issues. It has introduced new research methodologies into the challenge of security confidence-building in the region and played a leading role in the creation of second-track dialogue and the initiation of so-called one-and-a-half track regional security exchanges. New areas of research have been initiated in the areas of terrorism and transnational crime while, at the same time, a more traditional focus has been retained on the question of US strategic primacy and the changing role of alliances in the Asia-Pacific region. A shortcoming that needs to be addressed is the Centre's lack of specific academic expertise on China, Japan and India. We were fortunate in 2001, however, to attract Ron Huisken back to SDSC from Defence — he brings formidable knowledge of the region, the US alliance, and nuclear proliferation issues.

What Happened in Australia and SDSC's Response

From the early days of its creation, SDSC was at the forefront of developing new ideas for Australia's defence policy. Tom Millar, Bob O'Neill, Jol Langtry, Des Ball and Ross Babbage were instrumental in developing ideas for a more independent Australian defence policy, as well as detailed analyses of the geography (and defence strengths and vulnerabilities) of the north of Australia. Much of this work was path-breaking and ahead of the policy work being done at that time in Defence. I wish to record here my debt to these colleagues for the way in which their scholarship greatly assisted me when I wrote the *Review of Australia's Defence Capabilities* in 1985–86. Rarely for any form of academic endeavour, they had to begin with what amounted to a clean sheet of paper when considering planning for the defence of Australia. Their pioneering work, together with studies directed by Sir Arthur Tange in the Defence in the 1970s, made my task infinitely easier.

Most Australians these days accept only too readily the obvious statement in the 2000 Defence White Paper: 'At its most basic, Australia's strategic policy aims to prevent or defeat any armed attack on Australia. This is the bedrock of our security, and the most fundamental responsibility of government.'[6] In many ways, the history of the idea of the defence of Australia had its genesis in SDSC. That is little understood these days, including in Defence.

In 1990, Babbage published the major book *A Coast Too Long: Defending Australia Beyond the 1990s*, which was described by the Minister for Defence Kim Beazley at its launch as 'a unique contribution to the defence debate in Australia'.[7] Babbage also published a large monograph, *The Strategic Significance of Torres Strait*, which was prepared as a report by the Centre to the Minister for Defence. In the same year, Ball and Langtry edited the book *The Northern Territory in the Defence*

---

6   *Defence 2000: Our Future Defence Force* (Canberra: Defence Publishing Service, 2000), p. 29.
7   *Annual Report 1990* (Canberra: Research School of Pacific Studies, ANU), p. 26.

of Australia: Geography, History, Economy, Infrastructure, and Defence Presence, and the Centre published monographs on the management of weapons systems projects in Defence and the employment of air power in the defence of Australia. A major consultancy was undertaken for the Australian Army on the relevance of land forces in the defence of Australia. In 1991, a conference on 'Australia in Space' was also organised by SDSC. This was the first time that the civilian and military uses of space for Australian national purposes had been discussed by representatives from government, industry and academia.

SDSC is unusual in Australian academia in that it has undertaken policy-relevant research. With the agreement of ANU, I was adviser to two secretaries of Defence between 1991 and 2002. This involved giving high-level advice on strategic policy and force structure. It also involved preparing a consultancy report in 1995 on the future role and structure of the Papua New Guinea Defence Force.[8] Other members of the Centre, including Stewart Woodman and Mike Gilligan, assisted New Zealand, the Philippines and Thailand with developing their strategic policy and defence white papers.

A major external review of the Institute of Advanced Studies, presented in September 1995, referred to the impact that the Research School has had among policymakers, and singled out SDSC for its 'significant role in changing the orientation of Australia's defence thinking away from forward defence to one which emphasised a high level of self-reliance within an alliance framework'.[9] Throughout the 1990s, SDSC maintained its position as the leading academic authority in Australia on strategic and defence problems. With the change of government in 1996, the Centre moved quickly to remain relevant to the current debate and to meet the requirements of government policymakers. Yet, both then and now, SDSC has always been willing to push the debate beyond political or official pronouncements (this was especially the case when Senator Robert Hill was Minister for Defence between late 2001 and early 2006).

---

8   Paul Dibb & Rhondda Nicholas, *Restructuring the Papua New Guinea Defence Force: Strategic Analysis and Force Structure Principles for a Small State* (Canberra: Strategic and Defence Studies Centre, ANU, 1996).
9   *1995 Annual Report*, p. 78.

In the late 1990s, reductions in ANU financing of SDSC resulted in a move by Research School of Pacific and Asian Studies (RSPAS) to not give preference to Australian research priorities; there was to be no funding for Australian research. The *1997 Annual Report* of the Research School noted plaintively that research on Australian security was crucial to the Centre's work and to its influence in shaping official defence thinking. It asserted that '[r]esearch on regional security, from an Australian standpoint, would be meaningless without a detailed understanding of Australian defence'.[10]

In 1998 the Centre held a major international conference, 'Maintaining the Strategic Edge: the Defence of Australia in 2015', at Parliament House in Canberra. The conference attracted speakers from overseas, and three members of the Centre also presented papers. Despite the travails of ANU funding, SDSC continued its research on Australian defence policy, historical perspectives on current defence problems, Australian Defence Force (ADF) command arrangements and strategic and operational concepts, and Australia's maritime strategy. The corner was turned in 1999, such that the Centre could give increased attention to Australian defence.[11] Research concentrated on future strategic and defence planning, the Defence budget, implications of regional engagement for the Army, the need for local support to sustain Air Force operations, the implications of the East Timor deployment, an analysis of the current capabilities of the ADF, and a study of the Defence secretary's role in Defence policymaking.

The decade ended on a high note when I was asked by the Secretary of Defence and the Chief of the Defence Force (CDF) to be the co-leader of the internal consultation team within the Department of Defence that supported the government's public discussion process leading to the development of its Defence White Paper, which was released in December 2000. Myself and Air Vice-Marshal Brendan O'Loghlin visited 47 Defence establishments and had 64 meetings with over 6,000 members of the ADF in a period of only six weeks. Discussion of strategic issues highlighted the consensus view that the defence of Australia was the primary role of the ADF and there was also strong support for engagement in the region. The ADF's force 'hollowness' was

---

10   Director's section, *1997 Annual Report* (Canberra: Research School of Pacific and Asian Studies, ANU), p. 3.
11   *1999 Annual Report* (Canberra: Research School of Pacific and Asian Studies, ANU), p. 19.

of great concern and people were also critical of Defence's acquisition record. The feeling of 'change fatigue' was widespread. Members of the ADF were not convinced of the true savings from commercialisation and felt the process had gone too far and was undermining military capability. There was a strong feeling that Defence leadership was not adequately addressing personnel problems.

Ball's edited book *Maintaining the Strategic Edge: The Defence of Australia in 2015* was published by the Centre in 2000, and Horner completed his book *Making the Australian Defence Force* (volume 4 of *The Centenary History of Defence* series) in 2001, looking at the ADF's development as a joint force. Between 1998 and 2002, as an Army Reserve Colonel, Horner was the first head of the Army's Land Warfare Studies Centre while still being attached to SDSC.

The advent of a new Defence minister, Senator Hill, late in 2001 heralded the beginning of a wide-ranging — and sometimes destructive — debate for the next four years about whether the ADF should be primarily structured for the defence of Australia and regional contingencies or whether it should be a global expeditionary force, operating in a subordinate role to our US ally. SDSC played a key role in both the academic dialogue and the debate in the media. In this period, I wrote over 70,000 words on this subject, collected as Canberra Paper No. 161, *Essays on Australian Defence*.[12] It was launched in July 2006 by Allan Hawke, the new Chancellor of ANU and former Secretary of the Department of Defence, who in his speech described SDSC as 'a national asset'.

Military History at SDSC

The study of military history at SDSC has a long and proud tradition, beginning with O'Neill's seminal volumes on the official history of the Korean War. In the decade of the 1990s, Horner was recognised as Australia's leading military historian. He was made a full professor in 1999 (Professor of Australian Defence History) and was honoured with the appointment as the Official Historian of Australian Peacekeeping and Post-Cold War Operations.

---

12   Paul Dibb, *Essays on Australian Defence*, Canberra Paper No. 161 (Canberra: Strategic and Defence Studies Centre, ANU, 2006).

Despite this, we struggled to have Horner's academic qualities recognised by RSPAS when, as noted earlier, it faced serious budget problems in the mid-1990s and decided that academic priorities for Australian subjects should go to the back of the queue. This view was misguided on at least three grounds: first, strategic studies cannot be undertaken on a purely geographical basis; second, strategic and defence studies is based strongly in history; and third, SDSC has always supported history with a strong policy relevance — not military history for its own sake.

Horner, who took up the post of research officer in SDSC in September 1990, began by undertaking an analysis of Australian defence command and organisation. By the following year, he and Woodman had produced the major book *Reshaping the Australian Army: Challenges for the 1990s*, which was produced under a consultancy agreement with the Australian Army. In 1992 Horner was an historical adviser to Prime Minister Paul Keating during his visit to Papua New Guinea and published a book on the ADF in the Gulf War.[13] He was awarded a grant from the Australian Army for a history of Australian artillery (*The Gunners: A History of Australian Artillery*, published in 1995) and for a biography of Field Marshal Blamey (*Blamey: The Commander-in-Chief*, published in 1998), as well as from National Archives of Australia for a book on the War Cabinet (*Inside the War Cabinet: Directing Australia's War Effort, 1939–1945*, published in 1996). In 1994 Horner published *The Battles That Shaped Australia*, and was appointed editor of the Australian Army's history series.

In 1995 he was awarded a grant from the Department of Defence to undertake a study of Sir Frederick Shedden and his influence on defence policy, *Defence Supremo: Sir Frederick Shedden and the Making of Australian Defence Policy* (2000). He also wrote a book with Ball, *Breaking the Codes*. This is a formidable publishing achievement and Horner is to be warmly congratulated on his serious approach to military history and his outstanding productivity. I look forward to reading what I am sure will be a profoundly good official history of Australian peacekeeping.

13   *Annual Report 1992* (Canberra: Research School of Pacific Studies, ANU), p. 28.

Education and Training at SDSC

SDSC has a rather chequered history when it comes to education and training. This is because ANU policy towards postgraduate education has varied over time and the attitudes of some of our academic colleagues towards training has been one of disdain. External interest has also varied greatly; at present it is at a high point for both PhD and Master's education and we are hard pressed to keep up with demand. Yet, in the late 1990s, SDSC was told by Defence that it had no interest in postgraduate qualifications in strategic and defence studies, as distinct from short training courses and workshops.

In the early 1990s, the graduate program in strategic and defence studies, coordinated by Leszek Buszynski, had 12 students, including two from Indonesia, one from Malaysia, one from Singapore and two from New Zealand.[14] The program was refined under my direction to reflect greater focus on the region and to place more emphasis on Australian defence policy formulation. All members of the Centre were involved with the program and, in addition to Buszynski, specific courses were conducted by Woodman and Gilligan. In 1992 two doctoral students joined the Centre and this number grew to four by 1994.

Master's students increased to 20 in 1993 and stayed at that level for the following two years. In a major initiative, agreement was reached for a joint Master's program with the Singapore Armed Forces Training Institute to commence in 1995.[15] Woodman was appointed director of graduate studies at SDSC from early 1994 in succession to Buszynski, who took up an appointment at the International University of Japan. Woodman introduced a new unit entitled 'Defence Planning and Decision-making in the 1990s', which was an interactive unit requiring students to prepare for, and perform in, realistic policymaking situations. SDSC also began to conduct ad hoc courses for government departments and officers from foreign countries. By the late 1990s, however, we decided to terminate the Master's program because of dwindling numbers and Defence's lack of interest in providing financial support.

---

14 *Annual Report 1991*, p. 22.
15 *1994 Annual Report* (Canberra: Research School of Pacific and Asian Studies, ANU), p. 53.

A decision was made in June 2001 to bring Babbage back to SDSC as an adjunct professor to reinvigorate the Master's program in a major way. He did this very well, bringing great dynamism and energy to expanding the Master's program to operate in Australia as well as in a great many overseas nodes. Since then, Robert Ayson, the Director of Graduate Studies, and his highly competent team have rationalised the Master's program to focus essentially on teaching in Australia. The Centre expanded to teach about 90 Master's students and 12 PhD students. This is an all-time record for SDSC and it has made a major difference to our reputation as an academic centre of excellence in postgraduate studies. Under Professor Hugh White's guidance, we are now also delivering a major new training program to Defence's graduate trainees.

The State of SDSC Finances

SDSC finances have rarely been strong — we have always been a small organisation operating on a shoestring budget. From time to time, we have been able to raise external sources of funding, yet in my tenure as head I always considered that we needed to be careful because of the sensitive nature of the topics we research and the fact that it is all too easy to be seen to be beholden to the particular source of funds being offered. It is more prudent to politely refuse funding that could be perceived to be tainted, even if there are no obvious strings attached.

When I took over managing SDSC from Ball in 1991, ANU funding of the Centre was at its highest point, but I immediately took steps to broaden and strengthen its financial base. I approached the then Secretary of Defence, Tony Ayers, who undertook to double Defence's contribution to SDSC with a commitment to endeavour to maintain that level of funding for four years.[16] In 1995, Defence was providing just under half the Centre's funds.[17] This funding enabled us to appoint a new visiting fellow and it also placed the graduate program on a more sound financial base. Defence, the Ford Foundation, the New Zealand Ministry of Defence, British Aerospace Australia, Rockwell and Boeing all provided student scholarships and other support

---

16   *Annual Report 1992*, p. 28.
17   *1995 Annual Report*, p. 80.

for the graduate program.[18] The CSCAP program, whose Australian office is located in SDSC, was supported by grants from Foreign Affairs, Defence, Tenix Defence Systems, Raytheon International and Australian Defence Industries. I believe that our financial success in this regard reflected our academic performance in developing new ideas for regional security at a time of great international strategic change and in ensuring that the Centre's research was more policy relevant.

In 1994, SDSC was successful in gaining a National Priority Reserve Fund grant from the Department of Employment, Education and Training (now the Department of Education, Science and Training (DEST)) to fund research on security issues in north-east Asia, while substantial funds were also provided by the Department of Foreign Affairs and Trade for work on regional confidence-building measures.[19] This enabled us in 1994 and 1995 to engage four new staff members to concentrate on various aspects of regional security. Faced by restrictions on the availability of ANU funding, the Centre successfully expanded to an academic and research staff of 13, even though there were only three ANU-funded positions.[20]

The period from the early to the late 1990s was a time when half a dozen or more sources of external funding generated an increasingly robust academic research agenda for the Centre. This changed in the late 1990s as ANU funding was progressively cut and we subsequently also lost a major part of our Defence funding due to the creation of the Australian Strategic Policy Institute (ASPI) in 2001, as well as the fact that we had terminated our Master's program at the end of 1997 due to lack of interest in Canberra. We had to reduce our academic staff from seven to five and our support staff from six to four by the end of 1997.[21] Those of us who earned consultancy income were able to cushion the impact of these cuts on the Centre to some extent by subsidising academic endeavours (such as conferences) or making financial gifts to ANU that could then be reallocated according to donor priorities.

---

18 *1993 Annual Report* (Canberra: Research School of Pacific Studies, ANU), p. 28.
19 See *1993 Annual Report*, pp. 28–29 and *1995 Annual Report*, p. 81.
20 *1995 Annual Report*, p. 81.
21 *1996 Annual Report* (Canberra: Research School of Pacific and Asian Studies, ANU), p. 77.

As SDSC entered the new century, its finances improved once again, although they can hardly be described as robust. White has been successful in obtaining Defence funding for a new post-doctoral position, as well as PhD scholarships and the new Defence graduate training program mentioned earlier. Funding from ANU, however, continues to be tight. We are fortunate that visiting fellows such as Admiral Chris Barrie, Coral Bell, Richard Brabin-Smith, and the Hon. Derek Quigley provide their services to us free of charge. We are also fortunate to have the services of such experienced people as John McFarlane, Alan Stephens, Ross Thomas, Clive Williams and Derek Woolner. It is this mixture of academics and former senior Defence officers that gives SDSC its great depth of both theory and practice, despite the periodic vicissitudes of our financial position.

The Challenge of Managing SDSC

As a former senior Defence bureaucrat, I cannot say that I found my 1991 transition to managing SDSC particularly easy. There were three reasons for this: the first was the precarious nature of SDSC finances; the second was the denigratory attitude of some of our academic colleagues to the work of the Centre; and the third was my reaction to the overly bureaucratic decision-making processes at ANU.

The challenge of managing SDSC finances is perhaps best reflected in a story from my first day as head of the Centre. The previous day I had left my position as Deputy Secretary in the Department of Defence where I considered no expenditure under the sum of $20 million. I knew that the change would be painful — but little did I suspect that the first account I had to sign was $20 for an afternoon tea! The more serious challenge was that SDSC had little clout in the 1990s when it came to academic in-fighting for a share of the Research School's declining budget. The fact that we were a small Centre, as distinct from a full-blown academic department, counted against us — as did the nature, I suspect, of our academic work.

This brings me to the problem of negative attitudes towards SDSC. The nature of our work makes it distasteful to some of our colleagues and, they allege, not worthy of serious scholarship. Ball brought to my

attention some observations about the Centre by a visiting professor
of international relations to the then Director of our Research School
in 1988. This professor asserted that:

> the work of SDSC, or much of it, seems more apt for a military staff
> college than a university. It should be vigorously cut back. Strategic
> studies are an integral part of IR, and should be taught as such in
> a university ... the technical aspects of strategy are not fit meat for
> a university ... there is a large SDSC covering an area that is widely
> acknowledged internationally not to be one of intellectual innovation
> or growth now compared with the 1950s and 1960s.[22]

Lest it be thought that this attitude was merely ancient history, I met
the full force of academic prejudice almost as soon as I took over as
head of SDSC. Professor Andrew Mack and his colleague Richard
Higgott, both of the International Relations department within our
own Research School, referred to the staff of SDSC as 'bomb-fondlers'.
More seriously, they mounted a concerted effort to undermine the
Centre as, under my guidance, we became more focused on policy-
relevant research both in the area of Australian defence and regional
security. The problem was that, as a small Centre, we were a part of
the Division of Politics and International Relations and the divisional
convener was Mack. He effectively managed the finances of the division
as a whole and, hence, the allocation of money between International
Relations, the Department of Political and Social Change, and SDSC.
As I proceeded to raise more external funding (mainly from Defence),
Mack started cutting SDSC's share of the divisional budget and made
it clear he would be making more substantial cuts.

There was also a growing rift intellectually, with International
Relations moving away from the predominant realist paradigm under
professors Hedley Bull and J.D.B. Miller (as well as professorial fellows
Millar and O'Neill) to a greater focus on theoretical approaches.[23]
The differences also resulted in petty quarrels over accommodation
in the H.C. Coombs building. This rift culminated in the decision by

22   A.J.R. Groom, Professor of International Relations at the University of Kent at Canterbury;
Co-Director of the Centre for the Analysis of Conflict; and Vice-Chairman, British International
Studies Association, made these remarks in a paper entitled 'Some Observations on the
International Relations Department, Research School of Pacific Studies', 15 Aug. 1988 (personal
papers of Professor D.J. Ball).
23   Paul Dibb, 'On Relations Between SDSC and IR', submission to the Review of International
Relations at ANU, 14 May 1999 (personal papers of Professor P. Dibb).

the then Director of RSPAS in 1994 to move SDSC from the Division of Politics and International Relations to become a separate Centre administered in the Director's unit.[24] SDSC also moved out of the Coombs building to accommodation in Acton House. This, of course, only served to increase our sense of intellectual separation from the mainstream of RSPAS.

There can be no gainsaying of the void between policy-relevant work and 'pure academic work'. Some of our colleagues seem to continue to be not entirely comfortable with what SDSC does, although let me stress that this is not the stance of the current Director of RSPAS or his predecessor. And there can be no doubt that relationships with International Relations have greatly improved under the stewardship of Professor Chris Reus-Smit and his senior academic colleagues (one of whom happens to be one of my best friends). The hatchet should be well and truly buried by the time we move into shared accommodation in the new Hedley Bull Centre for World Politics in 2008.

The third issue I want to raise is the problem of management in a university bureaucracy. When I worked in Defence, I thought it was one of the most ponderous organisations conceivable. I was wrong; university bureaucracies 'take the cake'. I soon discovered that universities consist of interminable meetings, with little in the way of hard conclusions, and endless, suffocating pieces of bureaucratic paper. The fact is that most academics are not good (or comfortable) with administration or pushing through unpalatable decisions in committee. I recognise, of course, that the collegiate nature of the academic faculty requires decision by consultation and consensus. Yet I have to say that I found it surreal how colleagues on the Faculty Board and Strategy Committee refused to contemplate cuts to the Research School's research priorities as the ANU budget was being slashed by the government. The reasoning from some colleagues was 'the government dare not cut us any further'. It did, and eventually we had to excise a much-valued area of academic research, Pre-history and Quaternary Dating, and transfer it to another part of ANU. My experience in other university committees, such as the Board of the Institute of Advanced Studies, was of huge numbers of people

---

24   *1994 Annual Report*, p. 53.

engaged in endless discussion. I also found the university's hiring and firing regulations to be time-consuming, especially with regard to support staff.

I had clearly come from the wrong background, and I think some of my academic colleagues saw my impatience as resulting from what they perceived as me not being 'a real academic'. But Ball, who is indeed 'a real academic', experienced exactly the same frustrations when he managed SDSC. In 1991 he wrote some notes about leadership for a talk that he presented at the ANU conference 'Leadership in a Changing Context'. He stated very clearly that being head of SDSC was a frustrating and debilitating experience because ANU had overburdened itself with unnecessary and trivial administrative and bureaucratic practices and processes.[25] He found that he was spending about three quarters of the week working on the administrative and other duties expected of a head. Ball acknowledged that academics are, in general, neither good administrators nor good crisis managers. He observed that the criteria for academic success — good teaching and research abilities and a good publications record — are essentially irrelevant to the duties of administration and crisis management.[26]

There can be no doubt that there is much unnecessary bureaucracy and administrative minutiae in university life these days. Not all of this is the fault of ANU: the demands of DEST and the Australian Research Grants Committee to fill in endless forms are truly daunting. In the latter case, it was sufficient to put me off ever applying for an Australian Research Council grant.

We run the risk of forgetting that the prime purpose of the heads of departments and centres within ANU is to exercise leadership with respect to their disciplines in Australia. Only a head who has an active research program can command the intellectual respect of colleagues that is necessary to the exercise of academic leadership. As Ball observed, 'the Head of SDSC has a special responsibility to provide leadership with respect to issues of national importance'.[27]

---

25 Desmond Ball, 'Leadership: Expectations of Heads', notes prepared for a talk presented to a conference for university heads in the faculties and the institute on 'Leadership in a Changing Context', organised by the Centre for Educational Development and Academic Methods (CEDAM), ANU, Batemans Bay, 1–2 May 1991 (personal papers of Professor D.J. Ball).
26 Ball, 'Leadership: Expectations of Heads'.
27 Ball, 'Leadership: Expectations of Heads'.

ANU cannot be a cloister; basic research and freedom of academic inquiry are essential to the purpose of the university. However, they flourish best when tested against the issues of the real world, especially in our chosen field of endeavour — strategic and defence studies. The Secretary of Defence, Tony Ayers, argued this well when he wrote to the ANU Chancellor in July 1996:

> The Centre brings much credit to the University for its contribution to the understanding of defence matters in the Australian community and in our region ... The Centre's excellent reputation in the region has ensured continuing participation in its programs by officers and civilian defence planners from regional countries. This helps to promote a rational and disciplined approach to defence policymaking in neighbouring countries. From the perspective of Australian Defence personnel development, the Centre's courses, programs and publications have directly benefited Australian Defence Force officers and civilian staff.[28]

This does not mean that SDSC should only focus on practical defence policy issues. We must continue to be grounded in academic scholarship on the security of our region and the contending theories of strategic studies. But we should not fall into the trap of modish academic fashion: for example, the so-called 'War on Terror' is not the same existential threat to the survival of the nation-state as global nuclear war would have been between the Soviet Union and the United States.

One final thought concerns SDSC's position in a competitive world. The fact is that there is now a proliferation of well-funded new research organisations in Australia: they include ASPI, the Lowy Institute for International Policy, the International Security Program at the University of Sydney, as well as the Kokoda Foundation. SDSC does not receive such lavish private sector or government funding and, as illustrated above, constantly operates on a financial shoestring. But one of our comparative advantages is that we operate within the university system where there are no financial strings attached — we can be frank in what we say on any subject without fear of angering our sources of funding. Our other big competitive advantage is that we have the most experienced collection of senior academics (including four professors) and former senior military officers and Defence officials anywhere in Australia (including a former Chief of the Defence Force

---

28   *1996 Annual Report*, pp. 77, 78.

and three former deputy secretaries of Defence). This lends tremendous prestige to our publications program and to our postgraduate teaching — a prestige that is not comparable elsewhere in Australia. Yet we must not rest on our oars: SDSC successfully came through the difficult transition of the 1990s stronger than ever. I have every confidence that, under White's leadership, SDSC can look forward to a continuing bright future as Australia's leading academic centre in strategic and defence studies.

7

Researching History at SDSC

David Horner

The Strategic and Defence Studies Centre (SDSC) now has the largest number of academic staff working in the field of military and defence history in Australia, and this should not be surprising, because history has always been critical to the study of strategy. This was particularly the case when in earlier times strategy was seen as 'the art of the general', but continued to be the case when strategy became the concern of politicians and, with the advent of atomic weapons after 1945, nuclear scientists as well. The introduction of nuclear weapons led to a new academic discipline, namely strategic studies, the imperative of which was exemplified by Bernard Brodie's famous 1946 statement: 'Thus far the chief purpose of our military establishment has been to win wars. From now on its chief purpose must be to avert them. It can have almost no other useful purpose.'[1] But history remained central to the new discipline. It was no coincidence that the distinguished military historian, Sir Michael Howard, played an important role in the founding of the International Institute of Strategic Studies (IISS) in 1958. As the historian Brian Holden Reid explained, Howard 'consistently argued that those who wrote about nuclear strategy and studied history "talked more sense" than those

1   Bernard Brodie, *The Absolute Weapon: Nuclear Power and World Order* (New York: Harcourt, 1946), p. 76.

who had not'.[2] Howard stated that he was 'unrepentantly a historian and not a social scientist. I think in terms of analogies rather than theories, of process rather than structure, of politics as the realm of the contingent rather than of necessity.'[3]

Yet, despite its importance, history has always had an ambivalent place in the research interests of the SDSC. From the earliest days of the Centre, researchers in the field of strategic studies in Australia struggled to define it as an academic discipline or, more correctly, were involved in a continuous struggle to have strategic studies accepted as a distinct academic discipline in Australia. It was not the same as the discipline of international relations, with its antecedents of political science, although strategic studies had much in common with international relations. More broadly, the study of current defence policy could be seen as straight political science, but there was more to it than just politics. Some aspects of strategic studies drew on operational analysis. When it came to weapons systems, a background in science was important. Hence some leading practitioners of strategic studies have been scientists and, indeed, when the Cold War was at its height and nuclear issues were to the fore, nuclear scientists were important contributors to strategic studies. Strategic and defence studies (to give it a broader title) can also include the study of operational concepts and, in this regard, previous service in one of the branches of the military provides a good foundation.

As emphasised earlier, history is an important component of strategic and defence studies. The best way of understanding strategic and operational concepts is to examine how they have been used in past conflicts. The allied discipline of military studies (or military science) also draws heavily on history. To understand command structures, leadership, the problems of introducing new technology, the challenges of recruitment, the stress of combat, and the myriad other facets of military science, the best starting point is history.

In short, strategic and defence studies is multidisciplinary in nature, and this has been borne out by the background of SDSC's staff over the years. The first head of the Centre, Tom Millar, came from the field

2   Brian Holden Reid, 'The Legacy of Liddell Hart: The Contrasting Responses of Michael Howard and André Beaufre', *British Journal for Military History*, Vol. 1, No. 1, Oct. 2014.
3   Michael Howard, *Studies in War and Peace* (London: Temple Smith, 1970), p. 13.

of international relations; indeed he was member and later professorial fellow in The Australian National University's (ANU) Department of International Relations, a position he retained while head of SDSC. But he started his working life as a regular army officer, and his Master's thesis was in the field of military history — it was on the defence of the colony of Victoria.

The Centre's second head, Robert (Bob) O'Neill was also a former regular army officer. His first degree was in electrical engineering but he later switched to diplomatic and military history. His PhD thesis was a groundbreaking study of the German Army and the Nazi Party. The next head, Desmond Ball, was originally an economist who spent the whole of his working life in academia. His successor, Paul Dibb, was a geographer who had spent time in the Joint (later Defence) Intelligence Organisation and was a deputy secretary in the Defence department. A more recent head, Hugh White, undertook his first degree in philosophy; he too had been a deputy secretary in Defence. The present head, Brendan Taylor, was a political scientist and, like Ball, has spent his working life in academia.

The ambivalence towards history within SDSC comes from the fact that, for the work of the Centre, history is not an end in itself, but a vital tool to be used to underpin contemporary strategic and defence studies. If the Centre were to produce only studies of military or defence history — even if they made an outstanding contribution to understanding Australia's past — it would be accused of not engaging with current issues, which was the reason for its establishment. In short, the Centre needed to be seen as a relevant player in the broad area of contemporary strategic and defence studies.

From the early days, research in SDSC focused on three areas: Australian defence, regional security, and global security. The importance and priority of these issues changed over time. During the 1970s, global security (including the important work by Ball on nuclear strategy) and Australian defence were the most important. After the end of the Cold War, regional security assumed greater importance, and military history was not central to the work of the Centre. In its early days SDSC had a small staff (by 1974 it had grown to three academic staff — O'Neill, Ball and the distinguished journalist, Peter Hastings — and a research officer — Jol Langtry, a former Army colonel) and they needed to focus on the key strategic and defence issues of the day.

Much of the early work of SDSC members, however, had a basis in history. One example was Ball and Langtry's edited book, *Problems of Mobilisation in Defence of Australia* (1980).

Despite the focus on current issues, from its early days SDSC produced some important publications in the field of military or defence history. Millar's first major book as head of the Centre, *Australia's Defence*, published in 1965 with a second edition in 1969, was not a military history book, although it drew on the past to situate Australia's defence concerns in a historical setting. But after he stepped down as head and returned to International Relations, Millar published *Australia in Peace and War: External Relations Since 1788* (1978). This major study became a standard reference book for decades. A second edition was published in 1991.

When O'Neill became head in 1971, he had already made his mark as an outstanding historian. As noted earlier, his PhD thesis was published in 1966 to wide acclaim. His account of his battalion's service in South Vietnam, *Vietnam Task* (1968), was more than a standard battalion history and had thoughtful comments about the conduct of the war. In 1969 he published a biography of General Vo Nguyen Giap, the Vietnamese general who commanded the communist forces in the First Indo-China War and the Vietnam War.

In 1970, shortly before he took over at head of the Centre, O'Neill was appointed the official historian of Australia's involvement in the Korean War. This was a major and prestigious appointment. Australia had had only two previous official historians: Charles Bean for World War I and Gavin Long for World War II. O'Neill generally spent half of each day at the Australian War Memorial while he worked on the official history, and the other half at SDSC.

His official history, *Australia in the Korean War 1950–53*, was published in two volumes: *Strategy and Diplomacy* (1981) and *Combat Operations* (1985). This was the first time that an Australian official war history had a complete volume devoted to strategy and diplomacy, indicating that the reasons why Australia was involved and a discussion of the diplomacy were just as important as the actual combat operations.

At that time SDSC did not have PhD students of its own; they were officially part of International Relations, but in a practical sense they were part of SDSC. O'Neill's first PhD student, Neil Primrose, worked on a military history subject — 'Australian Naval Policy 1919–1942' — and his thesis was completed in 1974. Another early PhD student was Carlyle Thayer, whose thesis, 'The Origins of the National Liberation Front for the Liberation of South Viet-Nam', was completed in 1977.

I first met O'Neill when he was a lecturer in the faculty of Military Studies at the Royal Military College, Duntroon, and I was a cadet there. After graduation into the Infantry Corps I served in South Vietnam and then returned to my overriding interest — military history. In 1974–75 I completed an MA thesis at the University of New South Wales (UNSW) at Duntroon and, although O'Neill was then at ANU, he was one of my supervisors. He assisted me to get my MA thesis published as *Crisis of Command: Australian Generalship and the Japanese Threat, 1941–1943* by ANU Press in 1978. By that time I was undertaking my PhD thesis at ANU under O'Neill's supervision. It was probably due to his expert supervision that I was awarded the Crawford Prize for the thesis. It was subsequently published as *High Command: Australia and Allied Strategy, 1939–1945*, in 1982, just as O'Neill headed off to the IISS in London. The Army gave me the time to complete my MA and PhD, following which I returned to work in the Army for almost a decade.

With O'Neill's departure, SDSC had no specific expertise in military history, but it was still able to make a contribution in this field. In 1968 SDSC started publishing the Canberra Papers on Strategy and Defence. Over the next 38 years, the Centre published 164 monographs in the series and, quite rightly, their focus was on current strategic and defence issues. But many of them had a historical basis and at least 11 were primarily military or defence history. These were: W.A.C. Adie, *Chinese Military Thinking under Mao Tse-tung* (1972); Geoffrey Jukes, *The Development of Soviet Strategic Thinking Since 1945* (1972); D.M. Horner, *Australian Higher Command in the Vietnam War* (1986); J.C. Blaxland, *Organising an Army: The Australian Experience, 1957–1965* (1989); Nicola Baker, *More Than Little Heroes: Australian Army Air Liaison Officers in the Second World War* (1994); M.C.J. Welburn, *The Development of Australian Army Doctrine, 1945–1964* (1994); R.N. Bushby, *'Educating an Army': Australian Army Doctrinal Development and the Operational Experience in South Vietnam, 1965–72* (1998); R.W. Cable, *An Independent Command: Command and Control of the 1st Australian Task Force in Vietnam* (2000); Bob Breen, *Giving Peace a Chance: Operation Lagoon, Bougainville 1994* (2002);

Reuben R.E. Bowd, *A Basis for Victory: The Allied Geographical Section, 1942–1946* (2005); and Blair Tidey, *Forewarned Forearmed: Australian Specialist Intelligence Support in South Vietnam, 1966–1971* (2006). These history publications generally had some current policy relevance in that they particularly dealt with strategy, command, organisation and intelligence.

To a greater or lesser extent, I played a role in securing the publication of nine of these volumes. After I left the Regular Army in 1990 I returned to SDSC, initially as a research officer with responsibilities that included the publications program under Ball as the series general editor. In 1994, I transferred to the academic stream, but kept my responsibility for the publications program (still with Ball as general editor) through until about 2005.

From the time of my appointment at SDSC in 1990, I tried to publish defence history and current defence books alternatively, but over the next 15 years I was not as successful as I had hoped. During that time I published 18 books, but only four of them were on current defence issues and another was a compilation of history essays.[4] The 13 proper history books were: *Duty First: The Royal Australian Regiment in War and Peace* (1990), *General Vasey's War* (1992), *When the War Came to Australia: Memories of the Second World War* (1992, with Joanna Penglase), *The Gulf Commitment: The Australian Defence Force's First War* (1992), *The Battles that Shaped Australia: The Australian's Anniversary Essays* (1994), *The Gunners: A History of Australian Artillery* (1995), *Inside the War Cabinet: Directing Australia's War Effort, 1939–1945* (1996), *Breaking the Codes: Australia's KGB Network* (1998, with Desmond Ball), *Blamey: The Commander-in-Chief* (1998), *Defence Supremo: Sir Frederick Shedden and the Making of Australian Defence Policy* (2000), *The Second World War: The Pacific* (2002), *SAS: Phantoms of War, A History of the Australian Special Air Service* (2002), and *Strategic Command: General Sir John Wilton and Australia's Asian Wars* (2005).

Breaking the Codes, which I co-authored with Ball, was a groundbreaking study covering the development of signals intelligence in Australia, counterespionage, and the establishment of the Australian Security Intelligence Organisation (ASIO). It complemented Ball's earlier work

---

4   The Battles that Shaped Australia (Sydney: Allen & Unwin, 1994).

on signal intelligence, and consolidated SDSC as a source of expertise on intelligence and security — a crucial component of strategic and defence studies.

I admit that not all of these books fitted into the template of being history with a current policy relevance, but I believe most of them did. In general, my focus was on problems of higher command and the interface between the military and their political leaders — an issue that has continuing relevance. I was also interested in recent military operations, in the belief that research on them was necessary to understand the contemporary problems faced by the military.

During this time, I sought to interact with the wider military history community. In 1993 I became an inaugural member of the Australian Army Military History Advisory Committee. At my suggestion the Army established a history series, to be published by a commercial publisher, and I was editor of the series from 1994 to 2012, during which time the series published more than 40 books on Australian Army history. In 1994 I was appointed Chairman of the Armed Forces Working Party of the *Australian Dictionary of Biography* (ADB) and later succeeded O'Neill as the armed forces editor of the ADB.

This focus on military history did not win wide approval within the Research School of Pacific and Asian Studies at ANU. The first problem arose from the view of some 'mainstream' historians that military history was not a legitimate academic discipline. This misguided view might have been a hangover from the Vietnam War era, when many young historians were protesting against Australia's involvement in the war. The second problem was that the history I was writing was focused mainly on Australia. How could such research be accommodated within a school that was focused on Asia-Pacific affairs? This latter argument also applied to other members of SDSC who were focusing their research on contemporary problems of Australian defence. Such research fitted within the charter of SDSC, but did not seem to fit neatly into the School's focus on the Asia-Pacific. Following this line of argument, if a world-renowned scholar or practitioner such as, for example, Henry Kissinger or Sir Michael Howard had wished to take up an appointment at SDSC they would have been rejected because their focus was not specifically on the Asia-Pacific region. SDSC objected to the argument about the need to focus research specifically on the Asia-Pacific region, especially when it was applied as the means of deciding funding within the School, but without success.

The net result of this approach was that the School decided that my position in SDSC should not be funded; that is, I would no longer be employed by the School or ANU. Fortunately, the Chief of the Defence Force, Admiral Chris Barrie, decided to fund a chair of Australian Defence History at SDSC. I was appointed and took up the position in July 1999. The title of Professor of Australian Defence History was chosen to emphasise the fact that I would be concentrating on matters that would have direct relevance to Australian defence. I would be concerned not just with the analysis of past military battles, but with broader issues, such as the organisation of Defence, the relationship between Defence and the government, strategy, defence policy, operational concepts and intelligence.

Two of my books, published soon after taking up my appointment, were my biographies of Sir Frederick Shedden, the longest-serving Secretary of the Department of Defence; and General Sir John Wilton, Chairman of the Chief of Staff Committee during the Vietnam War and a major contributor to the process that eventually resulted in the formation of the position of Chief of the Defence Force. The Shedden book was commissioned by the Secretary of the Department of Defence, Tony Ayers, before Defence decided to fund my position. As Professor of Australian Defence History, I had other research projects that were focused more on contemporary issues, one example being my book *Making the Australian Defence Force* (2001), which explored why this joint structure was formed in the 1970s and how it developed through to 2000.

Throughout this period, I was the only member of SDSC writing consistently on defence history matters although, as I mentioned earlier, other members occasionally included some historical background in their works on current strategic and defence issues. My solo work was to come to an end during the first decade of the new century.

In 2002 I was engaged by the Australian War Memorial (AWM) to undertake a feasibility study into an official history of Australian peacekeeping. The Official Historian for Australia's involvement in South-East Asia conflicts, 1948–75, Dr Peter Edwards, was appointed in 1982, and, by 2002, the series was nearing completion. In 2004 the Federal Cabinet appointed me as Official Historian of Australian Peacekeeping and Post-Cold War Operations. Unfortunately, the government did not make any allocation of money for this project but, in collaboration with the Director of the AWM, Major General Steve Gower, we were able to put together a project with funds from Defence (which paid my salary), the AWM and the Australian Research Council (ARC). The project initially envisaged four volumes. The AWM provided two authors and a research assistant, and the ARC funds allowed ANU to employ an additional author and three research assistants. The new author was Bob Breen, who had served many years in the Australian Army, including as an operational analyst, and had written several books on operations in Vietnam and Somalia. He had just completed his PhD at SDSC under my supervision on Australian force projection in the 1980s and 1990s. His thesis was published as *Struggling for Self Reliance* (2008) as part of SDSC's Canberra Papers on Strategy and Defence.

SDSC now became the hub of a major history research project. Later the project was expanded to six volumes, including one on overseas emergency relief missions, funded by the Department of Defence. By its nature, an official history takes time. The official historian and their team are granted access to all relevant government records and there is no censorship. Each history involves research into a vast number of government records and much cross-checking to ensure that the record is accurate and will stand the test of time. The first volume, *The Official History of Australian Peacekeeping, Humanitarian and Post-Cold War Operations*; Volume II, *Australia and the 'New World Order'* (which I authored), was published in 2001. Volume III, *The Good International Citizen*, written by myself and John Connor (originally from the AWM and later at UNSW at the Australian Defence Force Academy), was published in 2014. As explained later, more volumes were in the pipeline.

Building on the reputation gained by working on the peacekeeping official history, and the Centre's expertise in intelligence and security (*Breaking the Codes*), in 2008 SDSC submitted a bid to research and write the official history of ASIO. The Centre was successful with its tender and the project began in 2009 with me as project manager and general editor. The Centre was able to engage two more staff: Rhys Crawley and John Blaxland.

Rhys Crawley, who was completing his PhD on the August Offensive at Gallipoli at UNSW at the Australian Defence Force Academy, was engaged as the research assistant. Later, when he received his PhD, he joined the academic staff. His PhD thesis was published as *Climax at Gallipoli: The Failure of the August Offensive*, in 2014.

Early in 2011 the ASIO project was joined by John Blaxland who had spent many years in the Australian Army as an intelligence officer and, more recently, as Defence Attaché in Bangkok. More particularly for the project, he had published several military history books, including *Signal Swift and Sure: A History of the Australian Army Corps of Signals 1947 to 1972* (1998), and *Strategic Cousins* (2006), based on his PhD thesis, which he undertook in Canada. While working on the ASIO project, he completed his book *The Australian Army: From Whitlam to Howard* (2014), and published an edited volume, *East Timor Intervention: A Retrospective in INTERFET* (2015).

Like the peacekeeping official history, the ASIO project required research into a vast number of official records, most of which remained classified. Also, like the peacekeeping official history, there was to be no censorship of the ASIO history, but the government reserved the right to prevent publication of matters that might be damaging to national security. I wrote the first volume, *The Spy Catchers, The Official History of ASIO 1949–1963*, which was published in 2014. The volume was joint winner of the Prime Minister's Literary Award for Australian History in 2015 and received the British St Ermin's Hotel Intelligence Book of the Year award for 2015. The second volume, *The Protest Years, The Official History of ASIO 1963–1975*, by Blaxland, was published in 2015. The third volume by Blaxland and Crawley, *The Secret Cold War, The Official History of ASIO 1976–1989*, is due to be published in 2016.

The two official history projects, therefore, provided SDSC with a group of military historians who began to establish SDSC as a major centre of military history research. Because these historians were recruited for the official history projects, they did not detract from the Centre's traditional research on contemporary strategic and defence issues. Indeed, they were able to add to them. For example, Blaxland contributed his expertise in South-East Asian affairs and Blaxland and Crawley contributed to the Centre's teaching expertise in the area of intelligence and security. This teaching complemented the Centre's long-standing expertise in this area that was developed through Ball's extensive publications over the previous three decades.

The official history projects were not the only reason for the expansion of SDSC's history expertise. In 2008 Daniel Marston was appointed a research fellow in a position funded by Defence. Employed to work on contemporary operational issues, he came with a solid military history background also, having completed his DPhil at Oxford University, where he was supervised by O'Neill, then Chichele Professor of the History of War, and then having worked at the Royal Military Academy, Sandhurst. When he left SDSC in 2009 he was succeeded by Garth Pratten (arrived September 2010), who was the original principal research officer for the peacekeeping official history and then, like Marston, was a lecturer at Sandhurst. He had recently published *Australian Battalion Commanders* (2009).

In 2011, SDSC won the contract to deliver the academic component of the Australian Command and Staff College (ACSC) and, at about the same time, as a result of restructuring within the ANU, the Centre was required to teach undergraduate courses as well as running the Master's program. As a result, SDSC was able to advertise for considerably more staff. The staff to teach at ACSC needed an understanding of military affairs and candidates with PhDs in military history were well placed to gain appointments to many of the positions.

Marston returned from the United States, where he had held the Ike Skelton Distinguished Chair of the Art of War at the US Army Command and General Staff College, and was appointed Professor of Military Studies in SDSC and Principal of the Military and Defence Studies Program at ACSC. Marston had already published several books on military history and, once back in SDSC, he published *The Indian Army and the End of the Raj* (2014). The book was runner-up for the Templer Medal in the United Kingdom.

Other new staff members of SDSC with military history expertise who arrived between mid-2011 and mid-2013 included Peter Dean, Jean Bou and Russell Glenn, all of whom had written military history books. Dean came from the University of Notre Dame, Sydney, and had recently published a biography of Lieutenant General Berryman, *The Architect of Victory* (2011). After joining SDSC he published three edited volumes, *Australia 1942: In the Shadow of War* (2013), *Australia 1943: The Liberation of New Guinea* (2014), and *Australia 1944–45: Victory in the Pacific* (2016). Bou was a research assistant for the peacekeeping official history, during which he was joint editor (along with Peter Londey, one of the official history's authors, and me) of *Australian Peacekeeping: Sixty Years in the Field* (2009). He is the author of *Light Horse: A History of Australia's Mounted Arm* (2010). After joining SDSC he published *The Australian Imperial Force* (2016) and *The Australian Imperial Force in Battle* (2016). Glenn was an officer in the US Army, including service in the 1991 Gulf War and, after leaving the US Army, was an analyst with the RAND Corporation. He was already the author of numerous books and book-length reports, the most recent being a 2012 study of the second Lebanon War in 2006. His PhD was in American history with a focus on military history. Once at SDSC he published *Rethinking Western Approaches to Counterinsurgency* (2015).

SDSC's expertise in military history was strengthened further in early 2012 when Professor Joan Beaumont joined the Centre. She transferred internally from within ANU, having previously been Dean of Arts and Social Sciences. Beaumont is an internationally recognised historian of Australia in two world wars, Australian defence and foreign policy, the history of prisoners of war and the memory and heritage of war. She already had an impressive list of publications and, in 2013, she published *Broken Nation: Australians in the Great War*, which was a joint winner of the Prime Minister's Literary Award for Australian History in 2014 and winner of other prestigious awards. In 2015 Beaumont was responsible for the participation of ANU in a joint international conference with the AWM to mark the 100th anniversary of the landing at Gallipoli.

The strength of military history at SDSC is driven home by that fact that in the three years 2013–15, SDSC staff members published 13 books on military, defence or intelligence history. These statistics do not include SDSC's publications on contemporary strategic and defence issues, where the Centre's strong record has been maintained. In 2015, SDSC had 21 academic staff (including three emeritus professors working in the Centre) and, of these, nine could be counted as a military, defence or war historian.

In addition to the impressive quality of its staff members, SDSC's PhD students have also undertaken research in military history. Recent successful PhD students have been Steven Paget, whose thesis was on interoperability between the Australian, British and US navies in naval gunfire support during the Korean, Vietnam and Gulf Wars, and who is now a lecturer at the United Kingdom Staff College; and Tristan Moss, whose PhD was on the experience and role of the Australian Army in Papua New Guinea in the postwar period.

By 2014 the Official History of Peacekeeping, Humanitarian and Post-Cold War Operations was languishing, primarily because the government did not allocate funds to the project and the original funds, cobbled together from disparate sources, had long since dried up. Only two of the planned six volumes had been published. Breen completed Volume V, *The Good Neighbour: Australian Peace Support Operations in the Pacific Islands 1980–2006*, in 2011, but it had not been cleared for publication by key government departments. By that time Breen had left SDSC and was Director–Deakin University Post-Graduate Qualifications and Academic Adviser at the Centre for Defence and Strategic Studies, Australian Defence College (although he remained an SDSC visiting fellow). The volume was finally cleared early in 2015 for publication in 2016.

Steven Bullard of the AWM (and an SDSC visiting fellow) completed Volume VI, *In Their Time of Need: Australian Overseas Emergency Relief Operations*, in 2015 with the expectation that it will be cleared and published in 2016. That left two volumes to be completed. Encouraged by Prime Minister Tony Abbott, Defence made funds available and a major short-term research project began in March 2015. Londey began Volume I, covering peacekeeping missions that began between 1947 and 1989, when he was working at the AWM, but when he moved to the Classics program in the College of Arts and Social Sciences at ANU, he did not have enough time to complete the volume. The funds provided by Defence enabled him to be released from teaching to complete a large part of the volume, and he was joined by Crawley, who had completed his work on the ASIO project, and became joint author of Volume I. The Defence funds also allowed Bou to be released from his teaching duties at the ACSC and he became the lead author of Volume IV, covering missions in Africa after 1992. He was assisted by Breen, Pratten, the official history's long-serving research assistant, Miesje de Vogel, and myself. Volumes I and IV were scheduled to be completed by March 2016. This would mean that Australia's fourth official history series, comprising six volumes, had been managed and substantially written by staff of the SDSC, with some parts written by AWM staff working under my direction. There was a certain symmetry that the second head of SDSC, Bob O'Neill, had written the second official history series (on the Korean War) 30 years earlier.

While conducting the 2002 feasibility study for an official history of Australian peacekeeping, I was convinced that work needed to begin on another official history, covering Australian military operations in Afghanistan and Iraq. Following the terrorist attacks in New York and Washington D.C. on 11 September 2001, the Australian Defence Force sent troops into Afghanistan in October 2001; the troops departed in 2002 but returned to Afghanistan in 2005 and served there until 2014 when the commitment formally ended (although some troops remained). In 2003, Australia joined with the United States and the United Kingdom in taking part in the invasion of Iraq. Most of the Australian force withdrew soon after the initial campaign, but forces returned to Iraq in 2005 and remained there until 2009. Based on my experience with the peacekeeping official history, I knew that, with every day that passed, the writing of the new official history would become more difficult and, due to the political sensitivity of many of the operations, the Australian public and indeed the troops themselves had no idea what operations had actually been conducted, or why.

After agitating for many years, in 2011 I persuaded the AWM to commission a feasibility study into an official history of Iraq and Afghanistan and, in 2012, I undertook the study. The AWM Council agreed with my conclusion that a history was feasible and should begin as soon as possible. Attempts to obtain government approval were delayed by two changes of government during 2013 — Prime Minister Julia Gillard was replaced by Kevin Rudd in mid-2013 and, in turn, he was defeated in a general election by the Coalition led by Tony Abbott. As noted earlier, in 2014 the Abbott Government agreed to provide funds to complete the peacekeeping official history, thus clearing the way for a decision in April 2015 to fund the official history of Australia's engagement in Iraq and Afghanistan.

When SDSC was established on a shoestring in 1966, some of its far-sighted proponents no doubt hoped that 50 years later it would have developed to become the leading centre of strategic and defence studies research in the Asia-Pacific region. This is now the case. The Centre's research focus has changed slightly, but still bears some resemblance to its original structure. Whereas previously it was built around three pillars of research — Australian defence, global security and regional security — in 2016 its research was focused on three 'clusters' — Australian defence, military studies and Asia-Pacific security. Military and defence history at SDSC stretches across these three research areas, providing a crucial underpinning for research into contemporary issues. Further, with substantial teaching responsibilities (which was not the case in 1966) history provides an ideal tool for introducing students to many of the key concepts of strategic and defence studies. SDSC is now a major centre for research in the field of military and defence history. This outcome could hardly have been imagined in 1966.

8

Same Questions, Different Organisation: SDSC's Fifth Decade

Hugh White

Like all stories, the story of the Strategic and Defence Studies Centre (SDSC) over the last decade of its now 50 years is one of continuity and change. Interestingly enough, the big continuities in SDSC's stories relate to its external setting — the aspects of the world around us that we study. The parallels between the intellectual challenges we face in analysing and explaining Australia's strategic setting and policy responses today resonate surprisingly closely with those that inspired our predecessors to establish SDSC in the first place. On the other hand, SDSC today operates in a radically changed institutional setting, and has had to change the way it operates as an organisation in fundamental ways, especially over the past decade, in order to survive and flourish. In this chapter I will explore these two aspects of SDSC's story, to help (I hope) deepen our understanding of where we are today and the directions we should take in future.

Studying Strategy

SDSC is an unusual organisation with an unusual role. Established as a policy think tank within a university, it has always looked two ways: to the world of scholarship in all its aspects, and to the world of policy and public debate. As we will see in the later section of this chapter, SDSC's role as a scholarly institution, especially in relation to teaching in all its forms, has changed since those early days, and especially in the past decade. But, in other respects, the idea of a university-based think tank occupying this Janus-like position is as fresh and relevant as it was 50 years ago, and for much the same reasons. To see this, we have to go back and look at the circumstances of SDSC's founding in 1966.

Our founder, Tom Millar, and his colleagues picked their moment well. When SDSC was launched, Australia was facing a revolution in strategic and defence policy. Our strategic environment was in the midst of a profound change that would shake the foundations of our postwar strategic policy and require a major reorientation of our defence policy. We can see from Millar's account of those times[1] how clearly he saw the scale of the defence-policy challenge that Australia faced, and how important a part SDSC played in helping Australia respond effectively to that challenge.

Millar expressed the challenge with characteristic clarity and force in the opening paragraph of the paper on 'Australia's Defence Needs' that he delivered to the Australian Institute of Political Science (AIPS) in 1964. His chapter in this volume tells us how his AIPS paper was his first contribution to the academic study and public discussion of defence policy, the beginning of his life's work in this field, and hence in a very real sense the seed from which SDSC grew. After opening the AIPS paper with Thomas Hobbes' words on 'covenants without swords', he explains what he intends to cover:

> I shall discuss the 'swords' which Australia needs to possess if those covenants are to have any meaning for us and upon which, in the last resort, we must rely. For our great and powerful American ally and our somewhat less powerful but still very important British friend are not

---

1   Chapter 2, this volume.

inevitably committed to the defence of our continent and people and way of life. The security of Australia is primarily and ultimately the responsibility of Australians.

No one here will question, I hope, the right and need of Australia to have defence forces of some kind. The questions are — what kind? How many? And how should they be armed, equipped and organised? In what situations should we be ready — or may we be forced — to commit them?[2]

Thus, in a dozen lines, Millar set out the core issues of Australian defence policy, and the core agenda for the Centre that he founded two years later. His AIPS paper led to him write *Australia's Defence*, published in 1965. It remains a bracing and stimulating read today. Something of its clarity, directness, foresight and contemporary relevance can be judged by simply opening the front cover of the first edition. There, on the dust wrapper flap, in bold letters, is the question 'Can Australia Defend Itself?' And, printed on the end papers, is perhaps the first public example of the now infamous 'concentric circles' map: the hemisphere centred on Darwin, with rings indicating distance.

But, what is even more striking, the lines quoted above might serve as an agenda for SDSC today. To see why, it is worth looking a little more closely at the strategic situation that Millar was responding to, and the questions it raised. By 1964 Australia's postwar defence policy was already under great strain, and the first steps were being taken to rethink Australia's defence posture and transform Australia's military forces. Australia faced new and unfamiliar regional security challenges after World War II. The Asia that emerged from war after 1945 had almost nothing in common with the Asia of 1939. After the Pacific War, Australia was a different country, too, harbouring deep-seated fears about our vulnerability to attack from Asia, the possibility of which was proven in 1942. In the decade after the war's end, decolonisation and the threat of communism made the region suddenly more complex, and threatening.

---

2   T.B. Millar, 'Australia's Defence Needs', in John Wilkes (ed.), *Australia's Defence and Foreign Policy* (Sydney: A&R/AIPS 1964), p. 69.

The posture that became known as 'forward defence' was a specific response to these fears. It focused Australia's defence policy on encouraging and supporting the United States and the United Kingdom to be committed to our region and deal with these new regional security concerns for us. Forward defence is often seen as a product of the 'imperial' or 'global' tendency in Australian defence policy. I think that is wrong. Forward defence sought to engage Australia's global allies directly in addressing Australia's regional and local security concerns here in Asia, and especially in South-East Asia. In fact, under forward defence, Canberra repudiated its modest postwar undertakings to deploy forces beyond our region in the event of a global crisis, in order to focus on supporting the United States and United Kingdom in our own backyard.

For a country that did not want to spend much on its armed forces, forward defence made a lot of sense, while it lasted. But forward defence only worked as long as our allies played along, and as long as Canberra could be confident that they would use their power to promote Australia's interests and objectives. As it happened, the second of these conditions was the first to go. In the early 1960s it was already evident that Australia could not take the support of our allies for granted. First it became clear that Washington would not support Australia in its opposition to Indonesian incorporation of West Papua, and might not be sympathetic if Canberra found itself drawn into conflict with its large and increasingly well-armed neighbour. Later, during Confrontation, it became clear that London did not share Australia's interests in trying to manage the crisis in such a way as to improve the chances of a stable long-term relationship with Jakarta. By 1964, in other words, we had come to realise that America was inclined to be much softer on Indonesia than we wanted, and Britain was somewhat tougher.

The implications are obvious, at least in hindsight. Australia needed to do more to build its capacity to defend the continent and protect its regional strategic interests in South-East Asia and the south-west Pacific. This process began in the years 1962–64, which is much earlier than most people think. Without declaring a change in policy, the government under Robert Menzies set about transforming the Australian Defence Force (ADF) into a force that would be much better placed to defend Australia and its regional interests from local threats through the conduct of independent operations and unaided by its allies.

The government bought a host of new equipment, including F–111, Mirage, C–130 and Caribou aircraft, Huey helicopters, *Oberon*-class submarines, guided missile destroyers, M–113 armoured personnel carriers, and introduced conscription. Defence spending increased sharply. In two years from late 1962 to early 1964, through three separate major statements to parliament, the Menzies Government undertook the most radical changes to Australia's military capabilities in the postwar era, and laid the foundations of a defence force able to defend Australia and protect regional interests without relying on allies.

But, as Millar's account of his experience writing his 1964 paper suggests, this major change in Australia's defence policy was undertaken with little public debate or even public awareness. There was almost no public discussion on the strategic rationale of what was a major reorientation of the nation's military posture. The government did not issue a white paper or provide sustained explanation of the rationale for Australia's changing defence policy. Few outside government felt inclined or qualified to comment and, as Millar explains, the government made little effort to inform those who sought to understand and explain what was going on. Public attention, therefore, ignored the underlying strategic rationale of Australia's new defence posture, and focused instead on more sensational issues, such as procurement problems with the F–111s and conscription. This was the strange situation — major strategic change and radical policy innovation without serious public discussion — in which the academic study of Australian strategic and defence policy issues was born.

And all this, of course, was before the major commitment of US and Australian ground forces to Vietnam. By one of those quirks so common in history, the high-water mark of forward defence in Australia's commitment to the war in Vietnam came after Australia had already started to abandon the strategic underpinnings of the policy. The debates sparked by Vietnam shaped much of the environment of SDSC's earlier years. As Bob O'Neill's account in Chapter 4 of this volume makes clear, the intensity and passion of those debates made the academic study of strategic policy challenging. But the magnitude of the issues that Vietnam unleashed made the need for well-informed, rigorous, impartial and dispassionate debate about defence and strategic policy more evident than ever.

For a start, in the latter half of the 1960s, Canberra's earlier reservations about forward defence were overtaken by the growing doubts of our allies. By the end of the decade both the United States and the United Kingdom had decided, for different reasons, that their strategic postures in our region were unsustainable. For Britain the constraints were primarily fiscal: successive financial crises meant that the United Kingdom simply could not afford to maintain strategically significant forces in our region. For America, the reasoning was more complex, but the implications seemed just as clear: henceforth the United States would not defend allies in conflicts that did not affect the wider strategic balance. As far as Australia's regional security was concerned, we were on our own.

At the same time, however, other less threatening changes were occurring in Australia's strategic environment. In 1965 Suharto had replaced Sukarno and, over the next few years, Indonesia began to change from a strategic liability into a net security asset for Australia's regional security. South-East Asia as a whole began to emerge from decades of crisis and evolve into a region of peace and development, symbolised and supported by the development of the Association of Southeast Asian Nations (ASEAN). In China in 1966, the launch of the Cultural Revolution seemed to herald an era of anarchic self-absorption but, by the early 1970s, the United States and Australia were able to establish good relations with China and to dispel, at least for a while, Australia's major security concerns.

Meanwhile détente between the United States and the Soviet Union seemed to some to offer a safer global strategic balance. All these developments made Australia feel safer. By the early 1970s, the era of forward defence was clearly over. The good news was that our region looked much less threatening than it had for many decades. The bad news was that our allies had made it clear that we would have to deal ourselves with whatever problem might remain. All this vindicated Millar's formulation of the key issues in Australian defence policy and reinforced the need for broader, better-informed public discussion of strategic and defence-policy questions.

Fortunately, the new challenges stimulated perhaps the most active and informed defence debate we have ever had. With SDSC in the vanguard, a well-informed, sophisticated and diverse academic and public debate developed in which the government began to participate. Coalition Defence ministers, including John Gorton and Malcolm Fraser, aired new strategic ideas in public in the late 1960s. In March 1972, the Liberal Government of William McMahon produced a discussion paper that confirmed Australia's strategic policy had to change. It made a blunt assessment: 'Australia would be prudent not to rest its security as directly or as heavily, as in its previous peacetime history, on the military power of a Western ally in Asia.'[3] And it drew the inescapable conclusion:

> Australia requires to have the military means to offset physical threats to its territory and to its maritime and other rights and interests in peacetime, and should there ever be an actual attack, to respond suitably and effectively, preferably in association with others, but, if need be, alone.[4]

These ideas were conclusively established as the foundations of a new defence policy in the 1976 White Paper on Defence, published by the government of Malcolm Fraser. It is a remarkable document. The first chapter explained in a few lines the revolutionary changes of the preceding decade, and concluded:

> The changes mentioned above ... constitute a fundamental transformation of the strategic circumstances that governed Australia's security throughout most of its history.[5]

A few pages later, under the heading 'Self-Reliance', the White Paper explained the implications of this transformation:

> A primary requirement arising from our findings is for increased self-reliance. In our contemporary circumstances we can no longer base our policy on the expectation that Australia's Navy or Army or Air Force will be sent abroad to fight as part of some other nation's forces and supported by it. We do not rule out an Australian contribution to operations elsewhere, if the requirement arose and we felt that our presence would be effective, and if our forces could be spared from their national tasks. But we believe that any operations are much more likely to be in our own neighbourhood than in some distant or forward theatre, and that our Armed Services would be conducting operations together as the Australian Defence Force.[6]

---

3   Department of Defence, *Australian Defence Review* (Canberra: AGPS 1972) p. 11.
4   Department of Defence, *Australian Defence Review* (1972), p. 5.
5   Commonwealth of Australia, *Australian Defence* (Canberra: AGPS, Nov. 1976), p. 2.
6   Commonwealth of Australia, *Australian Defence* (1976), p. 10.

More than a decade after his AIPS paper, Australian defence policy had caught up with where Millar began in 1964. Meanwhile, however, there was more to be done. The principle of defence self-reliance was one thing; the practical detailed implementation was another. Sir Arthur Tange assembled a remarkable group of people within Defence to work on the conceptual foundations of an Australian self-reliant defence policy and, as O'Neill made clear, SDSC played a leading role in expanding and promoting this debate beyond Defence. But progress was slow, and many logjams remained when Kim Beazley became Defence Minister in 1984. To break them, Beazley commissioned Paul Dibb to review Australia's defence capabilities, and then write a new white paper, *The Defence of Australia* (1987).

Paul was of course a member of SDSC, where he had spent some time in between periods of very successful service with Defence. His work on the review and the subsequent White Paper crowned two decades in which Australian defence policy underwent a revolution, and two decades during which SDSC was consistently at the forefront of Australian defence-policy debate and development, through the work of Millar, O'Neill, Des Ball, Ross Babbage, Peter Hastings, Jol Langtry and many others. Key collections like *The Defence of Australia: Fundamental New Aspects* (1976) and monographs like Babbage's *Rethinking Australia's Defence* (1980) made major contributions to the development of the policies that came to be enshrined in the 1987 White Paper, set new benchmarks for the quality and sophistication of contributions to the development of national strategic and defence policy from outside the bureaucracy, and laid the foundations for the academic study of Australia strategic and defence questions.

This period provides important pointers for SDSC's future. Since the early 1980s, SDSC's scholars have produced work of international standing in many areas, such as Ball's work on strategic nuclear and regional security issues. But the heart of SDSC's contribution to Australia has been the quality of its work on questions relating to Australian defence and strategic policy throughout the decades. It is worth pausing to consider why this should be so. Of course the careful, impartial study of public policy questions has long been seen as one of the key roles of universities in society, and this was clearly a key purpose in the decision to establish a national university in Canberra 70 years ago this year. But strategic and defence policy poses some specific and unusual challenges that make it especially important that it be subject to the kind of study and analysis that universities can provide, and why it is best undertaken in a specialist multidisciplinary centre like SDSC. First, defence policy is conceptually demanding. Because wars are relatively uncommon, and major conflicts less common still, there is little scope to take an empirical or practical approach to designing strategic policies and defence forces. Major strategic and defence decisions are taken without clear knowledge of the circumstances in which forces will be needed, and with little chance to learn from experience. Instead, there is not much alternative to building more or less elaborate conceptual frameworks to guide decision-making. The rigour, consistency and adequacy of such concepts is thus critical to the quality of the policy. Academic study provides an excellent opportunity to explore and test these frameworks.

Second, work on defence and strategic policy must draw on a number of diverse areas of expertise. It of course involves disciplines like international relations, Asian studies and history, but it must also draw on expertise in military technology, the conduct of military operations, the organisation of defence forces and the functioning of bureaucracies, and national fiscal affairs. These technical, and sometimes arcane, aspects of the discipline are why effective public debate engaging real defence policy issues is relatively rare. It also means there is often less contestability in defence policy, either within government or outside it, than there is in other areas of public policy. That makes the role of centres outside government that can command the needed expertise all the more important in ensuring that defence policy ideas are rigorously analysed and imaginatively challenged. One might say that the role of a centre like SDSC is to bring to bear on questions of strategic and defence policy the traditional strengths and virtues of scholarship: careful analysis of assumptions, stringent attention to conceptual foundations, rigorous testing of evidence, full documentation, and strict impartiality. These were the qualities that underpinned the success of SDSC's contribution to policy debates in the past, and which can guide us in thinking about SDSC's future.

Third, universities are uniquely placed to integrate policy-relevant research with the development of expertise through teaching. For much of its history, SDSC has offered Master's and PhD programs that have helped expand the range and depth of strategic and defence expertise in Australia. The close integration of policy-focused research and graduate teaching provides an ideal environment for the development of high-quality skills that can raise the calibre of people available to work on these issues in government, the media, industry, NGOs and academia.

The need for such a contribution is as great now as ever. Despite the achievements of SDSC and other contributors to informed policy debates, strategic and defence policy skills remain in short supply in government, and Defence remains one of the few areas of public policy in which governments do not have a wide range of well-informed sources of advice and fresh ideas to draw on outside the bureaucracy. Nor is the public debate nearly as well-nourished with well-informed, accessible, expert and impartial analysis of policy choices and issues as it needs to be.

This has become clearer than ever in recent years. In another of history's tricks, Australia's defence policy response to the strategic turmoil of the 1960s and early 1970s was not completed until the mid-1980s, only a few years before the end of the Cold War, which raised a new set of questions about the nature of Australia's strategic situation and defence needs. New tasks and roles for the ADF sprang up, making our forces busier than they have been since Vietnam. Globalisation has changed, at least for some, how we conceive our strategic interests. New regional dynamics in Asia have raised questions about the future international order among the region's great powers, with potentially immense implications for Australia's security. New technologies have raised questions about the future role and nature of armed forces, and the development of air and naval capabilities throughout Asia has eroded Australia's military technological edge that, even in the 1980s, we tended to take for granted. And important new security challenges have emerged in Australia's immediate neighbourhood.

Over the 1990s these new developments were met by a wave of official policy papers. Between December 1989 and December 2000, Australian governments issued a total of seven substantial strategic policy documents, compared to only two in each of the previous two decades. But the terrorist attacks of September 11 2001 (9/11) and the subsequent War on Terror have injected new elements into the defence debate and raised new and perplexing questions. Uncertainty remains about whether 9/11 does, as some have claimed, mark a new strategic epoch, or whether it will be seen in retrospect as a distraction from deeper tides in our strategic affairs that raise major long-term questions about Australia's future security.

It has certainly increased confusion and uncertainty within government and in the wider community about the roles of our armed forces and the capabilities they need to perform them. Those questions are a long way from being resolved. SDSC has been prominent in these debates for well over a decade, with both Dibb and Alan Dupont, for example, playing leading roles from different perspectives. The government itself has realised the need for a stronger public debate and new sources of fresh policy thinking, and has supported the development of new voices and fresh ideas through the establishment of organisations like the Australian Strategic Policy Institute (ASPI), and through continued support for SDSC and other academic centres working in the security area. The field is growing, with new think tanks, like the Lowy Institute for International Policy, and new academic centres on international security at the University of Sydney and the University of New South Wales among others.

The past decade has only seen these trends deepen and the questions they raise become more focused. Terrorism continues to pose a bewildering policy challenge to governments around the world, including Australia's. At the same time, the US-led global strategic order, which seemed so robust in the immediate post–Cold War decades, now faces serious challenges in the Middle East, Eastern Europe and, above all, in Asia. A decade ago, a serious Chinese challenge to US leadership in Asia remained a debatable possibility. Today it is a clear reality, with implications for strategic affairs throughout Asia, and for Australia's approach to the management of its alliances and regional relationships, and its defence needs. Successive governments have failed to address these issues effectively, with two Defence white papers — one in 2009 and another in 2013 — offering no clear answers to looming policy questions. It is less and less credible to assume that the policy settings that have served Australia so well for so long will continue to do so in future, but the outlines of a new policy approach have yet to be established. Moreover, in a situation that would have been familiar to Millar, public and even expert debate on the choices that Australia now faces has done little to help clarify future needs.

All this sets an exciting and challenging agenda for SDSC's sixth decade. It means that SDSC's commitment to policy-related research is as important as ever. One of the key tests of SDSC's success is whether its work engages key policy issues and contributes to informing choices about them. That has important implications for the way SDSC directs and evaluate its work, and its audiences. SDSC's prime audiences include not only other academics, but those outside the academy who engage with the same issues on which we work. It means that SDSC's primary audiences will tend to be Australian; while we will always want to be engaged in, and informed by, international debates and developments, the natural focus of the Centre's work should be issues that bear on Australia's policy choices and, perhaps most specifically, on those where Australia's policy choices are clearly shaped by our unique circumstances.

Some will wonder whether this focus does not make us less 'academic' or 'scholarly' than a university centre should be. This is a question that has hovered around SDSC for most of its 40 years, as the accounts of my predecessors in this volume show. I think we need to address it directly. Scholarship is not defined by subject matter, but by approach. SDSC should aim to bring the disciplines and strengths of scholarly research to bear on questions of strategic and defence policy, just as economists and medical researchers do in their fields. It is hard to imagine an area of national life in which the clarity of scholarship is more obviously needed. SDSC's task is to address strategic and defence policy issues with the clarity, rigour, detachment, imagination, ethical standards and impartiality that are the true marks of scholarship. That is why we are part of a university. At the same time, we need to respect, and acknowledge how our work draws on, the more traditional disciplines: history, international relations, Asian studies, political science and many more. That is why SDSC is part of the College of Asia and the Pacific (CAP), where we are privileged to be part of a remarkable community of scholars. From them we have a great deal to learn, and the Centre should make it a primary goal to contribute as much as it can to their work.

This reference to our institutional setting here at The Australian National University (ANU) offers a segue to the second big theme I cover in this chapter — the changes in SDSC's organisational and financial environment, and the associated and vital issue of our developing role in education.

Surviving and Teaching

When I came to SDSC in late 2004, the academic environment was changing in ways that affected the institutional setting, financial basis and, in significant ways, the academic identity of the Centre. In the next few pages I will sketch those changes, and the way SDSC responded to them. This will be a personal perspective, offering an account of how things appeared to me at the time, and (in places) in retrospect.

The forces that drove these changes came from outside SDSC and had their origins in shifts in the wider university, encompassing the way that ANU saw itself and was organised, and these in turn originated in larger shifts in the tertiary education sector nationally. These were forces beyond our control, in other words. Our task was to respond to them in ways that gave SDSC the best path to the future, and overall that seems to have worked out.

SDSC was born, and spent its first four decades, within the Research School of Pacific and Asian Studies (RSPAS, then the Research School of Pacific (RSPacS)), which was one of the series of research schools that together constituted the Institute of Advanced Studies — the core of ANU since its establishment in 1948. An annual Commonwealth block grant funded the schools to undertake research and to train researchers through PhD programs. The research schools did not teach undergraduate courses, and taught very little graduate coursework. Education at ANU was primarily conducted by the faculties, which were seen as separate from the research schools, and which were organised and funded like those in Australia's other universities.

The block grant that funded the research schools was unique in the Australian university sector, and it was what made ANU special. The grant provided freedom from teaching and allowed the research schools to focus on research, which, in turn, assured ANU its place as Australia's most renowned research university. As a component of RSPAS, SDSC had benefited enormously by working in this environment, and there can be no doubt that this was central to SDSC's success. But it was, in a sense, too good to last.

By 2004, the university's block grant was declining steadily as a result of a decision made some years before when the Australian Research Council (ARC) was established as the primary mechanism for funding university research in Australia. ANU was allowed to compete for research funding from the ARC on the condition that the block grant was slowly reduced. The simple problem with this funding model was that ARC grants funded specific research projects, but they did not pay for the staff, basic facilities and infrastructure necessary to build, maintain and develop an institution.

As the block grant declined, these basic costs came under great pressure, which was inevitably transmitted down through the research school hierarchy to the coalface in departments and centres. The impact for SDSC was amplified by a decline in other important sources of funding. The Ford Foundation, which had long provided very generous support to SDSC staff positions, moved to different funding models that were not applicable to SDSC, and Defence reduced its funding support.

None of this detracted from the quality or the quantity of SDSC's research output. The Centre continued to publish widely and it was among the highest performing elements of RSPAS across many forms of output. Its high national and international profile, and the importance of the issues on which it was working, was clear. Indeed, the establishment and success over the preceding three years of ASPI and the Lowy Institute as think tanks focusing on similar issues was proof of the demand for the kind of policy-focused scholarly work in the field that SDSC had done so much to pioneer in Australia. Moreover, some specific areas of SDSC's work were financially flourishing. It is appropriate to mention the work of SDSC's historian, Professor David Horner, whose major project producing the official history of Australian peacekeeping operations was supported by Defence and by a major grant.

But by 2004 it was becoming clear that SDSC's funding model was no longer financially sustainable. We were running a structural deficit as the salaries alone of the core SDSC staff exceeded the block grant allocation, and each year the deficits mounted in the form of a debt we owed to RSPAS, which would eventually have to be repaid. Small grants, like one provide by Boeing Australia for the library, helped, and no doubt defence companies could have been persuaded to offer much bigger grants, but there were real doubts about the wisdom of relying too heavily on such sources. There seemed little prospect that other avenues of large external grants could be found to support the basic costs of running the centre. It was clear that some of the handful of positions in the core SDSC staff would have to go, and without new sources of funding more cuts would follow. Without a new financial model SDSC was in danger of disappearing.

SDSC's institutional position was also fragile at this time. Some years before, as a result of financial squabbles fuelled by intellectual and personal differences, SDSC was separated from the International Relations department of RSPAS, in which it had been incubated, and was attached instead to what was called the 'Director's Section' of the School — a small and untidy collection of units that didn't fit anywhere else. We received generous support from the School Director, Professor Jim Fox, and his successor Professor Robin Jeffrey but we lacked a larger affiliation within the School, which resulted in a certain vulnerability.

In fact we were not alone at all. SDSC's problems were shared, in different ways and to different degrees, by most if not all parts of RSPAS and ANU, and the solution to our problem was framed by wider changes in RSPAS and the university as a whole that unfolded over the next few years. The key to these changes was a radical shift in the place of education at ANU, which was reflected in a major change in the university's organisational structure. Essentially, ANU became more like a 'normal' Australian university, relying increasingly on revenue earned through education, both from government funding of student places and from students' fees, as the primary funding source. This was reflected in the abolition of the old split between the block grant-funded research schools and the education-funded faculties. Across ANU, a series of six colleges was established by amalgamating faculties and research schools. In our case, after a long and at times difficult process, RSPAS was amalgamated with the Faculty of Asian Studies to create the new College of Asia and the Pacific in 2009.

Four schools were established within this new college, including the School of International, Political and Strategic Studies —later renamed the Coral Bell School of Asia-Pacific Affairs — which became SDSC's new institutional home. (Note, in what follows I will continue to refer to it as the Bell School although it did not actually take that name until 2015.) Apart from SDSC, the Bell School comprised International Relations, Political and Social Change, and the State, Society and Governance in Melanesia Program. The new organisational setting restored the previously close link between International Relations and SDSC. While this created some anxiety in SDSC about International Relations dominating the new school and marginalising SDSC, in the event, this has not happened.

Under its first director Professor Paul Hutchcroft, and then under his successor Professor Michael Wesley, the Bell School has proved to be a congenial and productive setting for SDSC and has provided an excellent foundation for SDSC's rapid development into new areas. Moreover, although the Bell School was from its outset conceived as a loose federation of units from divergent disciplines and with differing priorities, the School has allowed us to benefit greatly through closer scholarly, administrative, outreach and personal links with colleagues in the other disciplines.

A vital part of this success resulted from the Centre's move into the Hedley Bull Building in 2009, which coincided with the establishment of the new School. Indeed, the process of planning and development of the building played a significant role in drawing together the elements of what became the Bell School over the ensuing few years. Planning for the new building began in 2004, following a grant from the Commonwealth Government to establish the Asia-Pacific College of Diplomacy, which included funds for a new building to house it. Professor Chris Reus-Smit, the head of International Relations, saw the potential for this project to be a way to bring together physically and, he hoped, intellectually and even organisationally the various elements of RSPAS that were working on international relations, broadly conceived, and to use the new building to integrate the image and branding of that work. Reus-Smit made a point early on of inviting SDSC to be part of this venture, which offered many attractions. The Centre had returned to the H.C. Coombs Building after a peripatetic period in which it had moved several times between buildings that offered few amenities and nothing by way of branding or identity. The idea of moving into a new, purpose-built centre as a more permanent and identifiable home had a lot of appeal. Perhaps inevitably there were also some fears in SDSC that this might prove to be an empire-building exercise by International Relations, reviving some of the concerns that had led to the split in the 1990s. Nonetheless, it was clear that the opportunities that the project offered far outweighed the risks and so, even before the process leading to the establishment of CAP were fully underway, SDSC was on track to move into the new centre and thereby build new and closer links with other elements of RSPAS working in related areas.

Reus-Smit's formidable drive and lobbying skills ensured that major additional funds were provided to supplement the original grant, resulting in the construction of the fine new building that SDSC occupies today. An early decision was taken to use the building to break down barriers between the different units that would share it, by spreading people across floors. From SDSC's perspective the decision to name the building after Hedley Bull, whose work on strategic questions was such a notable part of his overall achievement, was a welcome reassurance that our interests and approaches would be respected and supported by whatever institutional evolutions occurred once the move to the new building took place. And, as the transition to CAP gained momentum, it became entirely natural that the units moving into the new building should constitute themselves as a new school within the new college.

Thus, by 2009, SDSC found a new home both organisationally and physically, as part of the new Bell School and within the Hedley Bull Centre, which made a big difference to SDSC's sense of itself. While these moves did not in themselves do anything to solve the underlying financial problems that beset the Centre, they did provide a setting in which solutions to those problems could be more readily be found — in the business of education. A key rationale for the amalgamation of faculties and research schools to create the new colleges at ANU was to elevate education at all tertiary levels to become core business for all parts of the university, including those like SDSC that had formerly seen it as definitely secondary to research functions. For SDSC, future financial viability could only come from building for ourselves a strong income stream from education, and that is indeed what has happened.

It is important to note first that SDSC had long been in the education business. SDSC had contributed to International Relations Master's program since the 1970s, with our own program starting in 1987. This was successfully run until 1997. It was a small-scale program, usually with no more than a dozen or so students at a time, but it achieved high standards and produced some notable alumni. It was not financially self-sustaining, however, and, as money become tighter in the late 1990s, it was decided that the program was unsustainable. A fresh start was made in 2001 when Dr Ross Babbage, a very significant former member of the Centre who had maintained close links throughout his varied career in government and industry, came forward with a proposal to ANU to establish a new Master's program in strategic studies on a different and more ambitious basis, aiming for more students and to deliver the program not just in Canberra but at 'nodes' across Australia, Asia and beyond.

Partly because of the clear financial risks involved in such a bold scheme it was decided to establish the Graduate Studies in Strategy and Defence (GSSD) program as separate from SDSC. It is a tribute to Babbage's formidable drive and entrepreneurial flair that the GSSD program was launched in 2002 with its first intake of students, and it quickly grew. This early success also owed a great deal to the gifted young scholars and teachers that joined the GSSD staff at the outset — most notably Rob Ayson and Brendan Taylor, both of whom have gone on to make a major contribution both within SDSC and beyond. By the time I arrived at SDSC in 2004, the GSSD program was already a well-established and thriving success, and it was clear that it had the potential to grow further and to become not just financially self-sustaining but a basis for SDSC's long-term financial well-being. To get to that point, however, it was necessary to learn the lessons of the early years and modify the model somewhat in the light of experience. In particular, it became clear after the first few years that the model of teaching at nodes outside Canberra was not cost-effective, and so the focus shifted to teaching on campus. Over the next few years the GSSD program expanded and developed in several ways, provided the resources to recruit additional young staff and did a great deal to revitalise and energise SDSC. The administrative distinction between GSSD and SDSC, always rather faint, became increasingly irrelevant and was erased altogether as part of the process of incorporating SDSC into the Bell School in 2009.

It is worth noting that part of the success of the new and much expanded Master's program over the past decade or so can be seen as a reflection of a couple of broader trends. One is the trend across tertiary education in Australia towards increasing demand for postgraduate qualifications of all kinds. This has transformed graduate education from something of a niche cottage industry to a full-scale and competitive business, which has required SDSC to pay careful attention to the quality of its course, teaching and student administration. Quite apart from the quality of our teaching staff, we have been fortunate to have had a series of exceptionally capable professional staff managing our program who have made a real difference to the quality of our students' experience, and hence our competitiveness in the marketplace. The second trend has been growing student interest in our subject matter — both in national security more broadly conceived, and in strategic studies more specifically. It would be fair to say that Babbage's original conception recognised these trends, and the success of the Master's program as it has evolved over the past 15 years owes a great deal to the way those trends have been harnessed to build a viable long-term business.

Another useful element of SDSC's business development over this time was the expanded provision of short courses, especially to Defence. The most important of these was the three-day Strategic Policy seminar that SDSC provided to Defence's Strategy Division for delivery to members of the Defence Graduate Program. These seminars have been run, in different forms, for a decade, allowing SDSC to contribute to the development of policy expertise in Defence, as well as providing a valuable source of income.

These initiatives have contributed to the stabilisation of SDSC's financial position, and set it on a sustainable trajectory. Nonetheless, the Centre has continued to seek opportunities to expand its business and build a stronger foundation for future growth and development. One possibility emerged in 2008, when the newly elected government under Kevin Rudd decided to establish a National Security College (NSC), with a strong expectation that it would be located at ANU. The NSC was to be very generously funded and supported, and the question naturally arose whether SDSC should bid to take it on. In the end we decided not to, for several reasons. These included questions about its focus on 'national security' broadly conceived, the focus of its business model on professional and executive development courses, its governance arrangements and relationship to government, and the implications for SDSC's existing programs, brand and identity, which would risk being swamped by the new entity. The NSC was established successfully at ANU in connection with the Crawford

School, and SDSC cooperated closely in that process, with the result that NSC and SDSC now enjoy an excellent cooperative and healthily competitive relationship.

In 2011, a new and very different kind of major development opportunity arose, when the Australian Defence College (ADC) requested tenders for the provision of the academic element of the Australian Command and Staff College (ACSC) course delivered at its Weston campus. At first we approached this prospect with some caution. Over the years a number of institutions had undertaken this task in various different ways, but none had, so far as one could see, proven very satisfactory either to Defence or to the academic institutions involved. After some wary preliminary exploration, we decided that the ADC was serious about developing a new academic program based on a much more robust relationship with their academic partners. This included a decision by ADC to enter a 10-year contract with their new academic partner. This made a big difference to our thinking. We were determined not to take the job on unless we could do it very well and unless we could do it in a way that supported and enhanced, rather than detracted from, our established identity and activities. That meant we could only take it on if we could expand our staff significantly to accommodate the extra workload, and we could only do that if we could sign a long-term contract like the one ADC were prepared to offer. The prospect of a 10-year contract made that possible. We were also very impressed by the ADC's genuine openness to fresh ideas about how best to structure and deliver the academic element of the ACSC course. We had some strong and somewhat novel ideas about this, and were reassured that a tender based on those ideas would receive a fair hearing. So we decided to bid.

We were immensely fortunate that our colleague Stephan Frühling took charge of the Centre's bid for the ACSC contract. He developed an original, innovative and detailed course, a fleshed-out plan to deliver it, and a robust costing. Our bid was based on the provision to ACSC course members of a tailored ANU Master's degree program — the Master of Military Studies — within the 12-month duration of the ACSC course and alongside other elements of the ACSC program. This was a formidable undertaking. There is no space here for a full account of how it was done: suffice to say that the bid was successful and, in February 2012, and with some caution, SDSC entered into a 10-year contract with ADC. Within just a few months of the contract being signed, we began to deliver the program at Weston to some 160 course members. Frühling's effort in developing the bid in 2011 was eclipsed only by his achievement as the program's acting Director of Studies in 2012 as the program was implemented and bedded down. Since then, the program has evolved and flourished under the leadership of Professor Daniel Marston, who took over as Director of Studies in 2013.

The ACSC contract has made a significant difference to SDSC in many ways. Financially, it has transformed our business and strengthened our fiscal foundations. Academically and intellectually, it has deepened our engagement with core issues of strategic and defence policy, military operations and history, and defence administration — issues that have always been important to SDSC. Institutionally, it has consolidated our involvement with the real policy questions that confront Australia, as SDSC has always sought to do, and has strengthened our links with Defence and the ADF — including with the cohorts of rising officers who will lead the ADF in decades to come. And, perhaps above all, it has allowed us to hire a number of younger scholars who are doing a lot to shape the SDSC of the future.

Of course all this has happened under the leadership of my successor as head of the Centre, Brendan Taylor, who took over from me in late 2011, just as the ACSC contract was finalised. He deserves the credit for guiding SDSC through a remarkable period of expansion and development. During his time as head, the Centre's PhD program has been strengthened and expanded, and we have launched a remarkably successful undergraduate program. This is in many ways a new departure that is as significant and valuable to SDSC as the contract with the ADC, and it reflects and takes full advantage of those fundamental changes in organisation and outlook that I mentioned at the start of this section.

All this makes one rather optimistic about the future for SDSC in its second half century. Both the continuities and the changes that SDSC has seen over the past decade make me confident that a centre like SDSC has a big role to play, and that SDSC today is well placed to play it.

Plate 43 The SDSC team, May 2006

Plate 44 Professor Paul Dibb, the Hon. Kim Beazley, 2005

Plate 45 David Horner, Robert Ayson, Christopher Michaelsen, visiting guest Professor Reddy, Professor Ross Babbage, Brendan Taylor, John McFarlane

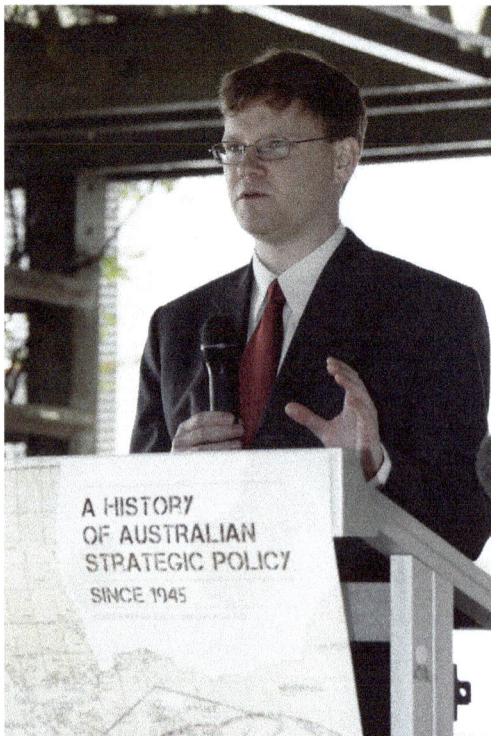

Plate 46 Dr Stephan Frühling at the launch of his book *A History of Australian Strategic Policy since 1945*, 2009

Plate 47 Professor Hugh White teaching a GSSD class, c. 2010

Plate 48 Professor the Hon. Gareth Evans AC QC, Professor Desmond Ball, Vice-Chancellor Ian Young, Professor Andrew Macintyre at the awarding of The Australian National University's Peter Baume Award to Desmond Ball, 2013. Photo by Charlie White.

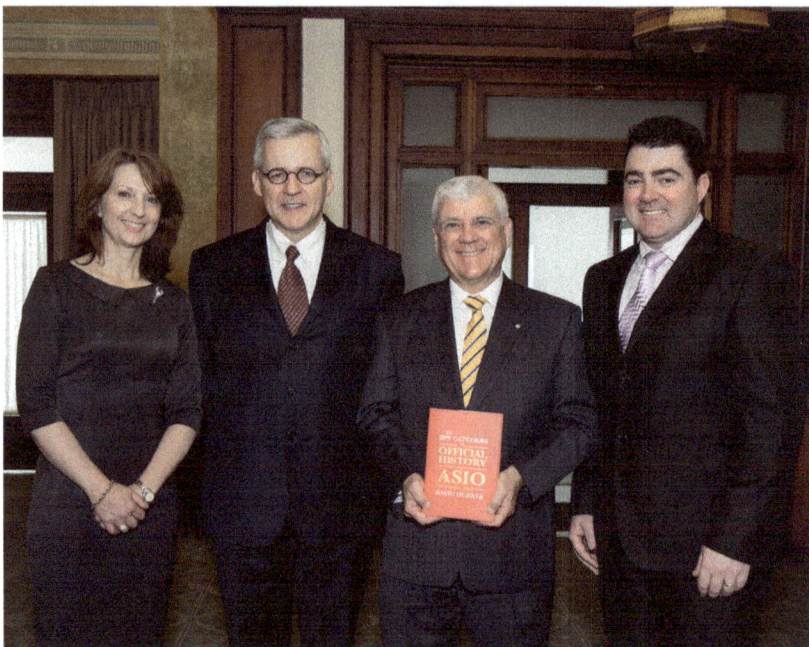

Plate 49 Lisa McKibbin, Dr John Blaxland, Professor David Horner
and Dr Rhys Crawley at the launch of *The Spy Catchers: The Official History
of ASIO 1949–1963*, 2014

Plate 50 Professor Desmond Ball, His Excellency, General the
Honourable Peter Cosgrove AK MC (Retd), at the investiture
of Ball's Order of Australia (AO), 2014

Plate 51 Professor Hugh White, His Excellency, General the Honourable Peter Cosgrove AK MC (Retd), at the investiture of Ball's Order of Australia (AO), 2014

Plate 52 Student with Chief of the Defence Force Air Marshal Mark Binskin at graduation for the Australian Command and Staff College, 2014

Plate 53 Graduating class of 2014 from the Australian Command and Staff College, 2014

Plate 54 Graduation address delivered by Chief of the Defence Force Air Marshal Mark Binskin, graduation for the Australian Command and Staff College, 2014

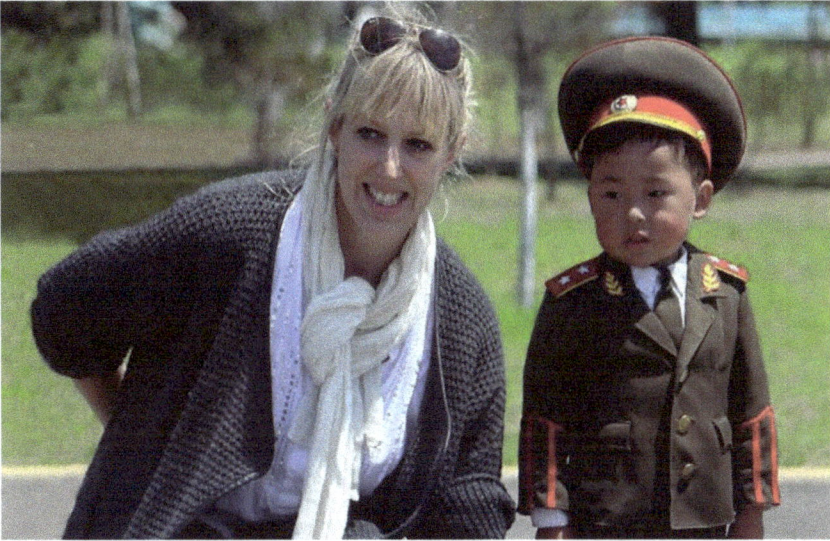

Plate 55 Dr Emma Campbell with young North Korean boy, North Korea, 2014

Plate 56 Professor Joan Beaumont with Prime Minister
Tony Abbott, receiving the 2014 Prime Minister's
Literary Award for Australian History, 2014

Plate 57 Professor Brendan Taylor, head of SDSC, at the launch of Graduate Research and Development Network on Asian Security (GRADNAS), 2015

Plate 58 Professor Evelyn Goh, 2015

Plate 59 Dr James Curran, Dr Andrew Carr and Dr John Blaxland, seminar on Australian Foreign Policy, 2015

Plate 60 General David Petraeus and Professor Daniel Marston, during class at the Australian Command and Staff College, 2015

Plate 61 Military and Defence Studies Program team, Australian Command and Staff College, Kathryn Brett, Daniel Marston, Belinda Corujo, Andrew Frain, Jack Bowers, Tamara Leahy, 2015

Plate 62 Visiting Fellow Dr Richard Brabin-Smith, Dr Amy King, GRADNAS launch, 2015

Plate 63 PhD student Ristian Atriandi Supriyanto, GRADNAS launch, 2015

Plate 64 Dr John Blaxland, GRADNAS launch, 2015

Plate 65 Dr Greg Raymond, GRADNAS launch, 2015

Plate 66 Attending scholars and graduate students
at launch of GRADNAS, 2015

Plate 67 Dr Stephan Frühling, visiting research fellow,
NATO Defence College, 2015

Plate 68 Former head of ASIO David Irvine with SDSC students
at 'Wine, Cheese and Wisdom' seminar, 2015

Plate 69 Professor Joan Beaumont, Ambon Island, Indonesia, c. 2015

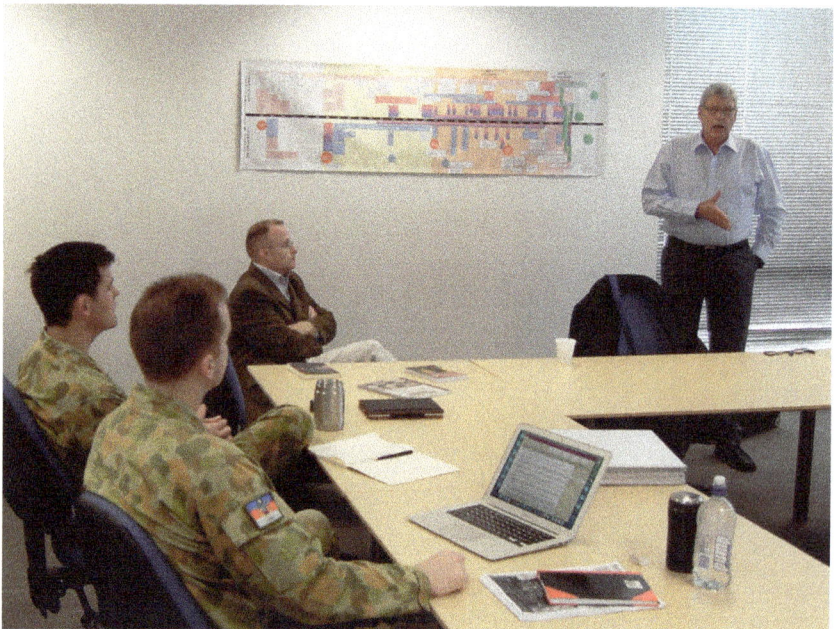

Plate 70 Professor Daniel Marston listens to a guest lecturer for his 'Art of War' class, Australian Command and Staff College, 2015

9

SDSC at 50: Towards a New Golden Age

Brendan Taylor

It was an afternoon in mid-2011 and I was working on the second floor of the Menzies Library — a habit that I had developed as a Master's student at the Strategic and Defence Studies Centre (SDSC) in the late 1990s. I was compulsively checking my email — a less positive habit that I also acquired in recent years — when a note came through from the head of Centre, Hugh White, asking whether I was free for a chat over coffee. Spending time in conversation with Hugh is one of life's great pleasures, and so I swiftly obliged and made my way over to The Gods Cafe on the ground floor of the Hedley Bull Centre.

The content of our conversation took me by surprise. Hugh had recently announced to the staff of SDSC that he was stepping down as head of the Centre after a stellar seven-year tenure and he was now in the process of canvassing potential applicants for the top job. He thought I should throw my hat into the ring. I was naturally flattered, but swiftly declined any interest. It was a job, I said to Hugh, that would perhaps tempt me a decade from now, but that I felt was too early to take on at that juncture in my career. In truth, the shoes of Hugh and his four predecessors were formidable ones to fill — too large for a mere 39-year-old, I felt at the time. Hugh, in his characteristically gentlemanly way, graciously accepted that decision but encouraged me to think the possibility over.

As with most major decisions in my academic career, I recounted this conversation to my mentor, Paul Dibb, who quickly indicated that I had an obligation to support Hugh's request. Out of respect to Paul, I thus informed Hugh that I would apply. To this day, I'm uncertain as to whether Hugh and Paul were in discussions on the next head of the Centre. Paul was certainly influential in attracting Hugh to the SDSC and I suspect that they probably had. Deep down, though, I still felt far from ready to take on such a nationally important and internationally prominent post, and my initial instinct was to go through the motions if I even made it to the interview stage.

As fate would have it, however, in the period immediately following these conversations with Hugh and Paul, I was in Singapore at a conference that was also being attended by former Chief of the Australian Defence Force and now SDSC visiting fellow Admiral Chris Barrie. Ever perceptive, Chris could sense my half-heartedness regarding the application and encouraged me to quickly think otherwise. He rightfully pointed out that poor performances in job selection processes are often remembered and that these can do terminal damage to one's career prospects over the longer term. I thus gave it my all at the September 2011 job interview and departmental seminar. The rest is history.

I still clearly remember the first time I walked the corridors of SDSC in February 1997 (when the Centre was located in what has since become the decidedly more trendy 'New Acton' on the outskirts of The Australian National University (ANU) campus) and my feeling of awe and slight intimidation on seeing the names of such giants in the field of Australian strategic and defence studies as Des Ball, Coral Bell, Paul Dibb and David Horner on the office doors. That feeling of inspiration and intimidation is one that never really goes away, especially when one walks past the pictures of the former heads of Centre that hang on the wall just outside my office today. It's difficult, in my own case anyway, not to feel like the 'odd one out'. Hence, if the 1990s constituted a period of 'difficult transition' for the Centre, as Dibb characterises it in his contribution to this volume, taking over from White as head of Centre in late 2011 was a transition that I found immensely difficult personally at such a relatively young age. Yet it was also a transition made easier by the support of those who had been most central to initiating it — Hugh, Paul and Chris — who made sure I had plenty of space to start making my own imprint on the Centre but who were also steadfast in their support and in their willingness to provide counsel when asked for it. I also sought the advice early on of my doctoral supervisor and another mentor, Des Ball, which as always was immensely helpful. Perhaps most importantly, it was also a transition made easier by having such a remarkably collegial team of colleagues. That collegiality is a theme to which I'll return in this chapter's conclusion.

White's headship was one which accomplished a great deal and which left the Centre in remarkably good shape. He managed to pull the SDSC out of some of its darkest financial days — at one point, from memory, we had run up a deficit of approximately $750,000 — to the point where we were once again 'in the black'. White was centrally involved in the resurrection of the Graduate Studies in Strategy and Defence (GSSD) program in the 2000s, which sprung to life as a result of Ross Babbage's entrepreneurial flair and which then thrived under the directorship of Rob Ayson. White and Ayson were a formidable combination. Respected scholars in their own right, they were both passionate educators and the period working with them through the mid-2000s, along with Stephan Frühling, Ron Huisken and our wonderful team of administrators, was undoubtedly the most enjoyable of my career to date — although Stephan still reminds me how hard we worked and how much marking we had to do!

In the fluid organisational environment of the ANU, however, White correctly recognised that the Centre needed to expand beyond its historically steady state of six to eight academic staff members in order to survive as an independent academic unit. By 2011, the Centre had become part of the newly formed School of International, Political and Strategic Studies (or IPS), which was an amalgam of similarly sized and minded units intent on maintaining an independent identity and, more importantly, budgetary autonomy. The inaugural Director of the School, Professor Paul Hutchcroft, understood and supported this desire and gave life to it by introducing an ingenious 'federated' organisational structure within the new school. SDSC was fortunate to have in Paul Hutchcroft such a strong supporter, which has enabled and facilitated the remarkable growth of the Centre over the past half decade. Such strong support has continued and deepened under its current Director, Professor Michael Wesley, who rebranded the School in early 2015, naming it after one of SDSC's own — the late Coral Bell.

The Centre has tripled in size over the past half decade, to the point where it is now home to more than 20 academic staff. In his contribution to this volume, Tom Millar marvelled during the early 1990s at the size of the SDSC oak that had sprung from the seed he planted in the mid-1960s. If only he could see the Centre today!

In addition to the steady income provided by the GSSD program, the Centre's remarkable growth has come about as a result of the expansion of its activities in two key areas. The first of these is the SDSC Security Studies program at the undergraduate level, which has burgeoned beyond all expectations.

The idea of an undergraduate program in security studies was not altogether new. During the early 2000s, when Babbage was (re)establishing the GSSD, he worked closely with Ball and the then Director of the Faculty of Asian Studies Tony Milner to establish a parallel undergraduate program based at the Faculty, but which could over time serve as a direct feeder into its SDSC graduate counterpart. The qualification associated with this undergraduate program was the rather awkwardly named Bachelor of Security Analysis (Asia-Pacific) — a title derived largely out of deference to colleagues across campus in the then Faculty of Arts and Social Sciences, who saw the new qualification as a potential competitor to their undergraduate offerings in the field of international relations.

Despite the new degree's institutional base in the Faculty of Asian Studies, SDSC continued to be centrally involved in its planning and delivery. SDSC staff frequently appeared as guest lecturers, while our doctoral students provided the bulk of the program's tutorial staff. Milner's successor, Kent Anderson, recognised this by the late 2000s and, both for the good of the program and as part of a restructuring of the Faculty into a reconstituted School of Culture, History and Language, he negotiated with White for the transfer of the program to SDSC. With this transfer, the degree was rebranded as the slightly more palatable 'Bachelor of Asia-Pacific Security', which continued to keep our colleagues across campus happy. In recent times, personnel changes and a shift in direction to a decidedly more quantitative methodological focus have eased pressures on that front. As such, it became possible for SDSC to re-brand its undergraduate program again in 2014 as an even more marketable 'Bachelor of International Security Studies'.

A second transformative event occurred in 2011 when the Centre won a substantial 10-year contract to deliver a graduate program to approximately 200 mid-level officers at the Australian Command and Staff College (ACSC). The successful bid was spearheaded by Frühling and it is no exaggeration to say that, without his leadership and entrepreneurship in winning this bid, the SDSC may well have succumbed to the restructuring processes that were occurring at ANU at that time. Instead, the SDSC almost overnight came to be regarded as a 'jewel in the crown' of the University (to borrow terminology used on occasion by its Chancellor and longstanding supporter of the Centre, Gareth Evans).

The early days of delivering upon this contract were demanding to say the least. I'd received some early advice that the biggest challenge of all would be our dealings with the military. In reality, nothing could have been further from the truth. I had had minimal, but generally positive, prior dealings with the Australian Defence Force (ADF). Consistent with this, SDSC's engagement with the leadership of the ACSC and the larger Australian Defence College has been extremely smooth during the first five years of this contract and I continue to be struck by the professionalism of the staff and course members there.

Yet the volume of work that needed to be done in order to match those high standards was considerable for a still-relatively small Centre, much of it falling onto the shoulders of Frühling (who acted as the inaugural Director of the program) and his dedicated deputy, Garth Pratten. Sensing my concern about the effect of this increased workload on the Centre, White quietly reassured me that things would settle down once we got through the first year. As I'm sure Stephan and Garth will attest, that first year seemed a very long one and I subsequently ensured that they both received well-earned sabbaticals in its aftermath as some form of compensation for their herculean efforts.

The upside of both the ACSC contract and the burgeoning undergraduate program was that they allowed the Centre to triple in size and to bring on a number of key new appointments.

Foremost amongst these was Daniel Marston who, by competitive process, won the position of Professor in Military Studies and Principal of the ACSC program, which subsequently came to be known formally as the Military and Defence Studies Program (MDSP). Marston is an Oxford graduate and protégé of Bob O'Neill. I'd got to know him a little when he succeeded me as the Centre's Defence-funded postdoctoral fellow during the mid-2000s — a position that he held for a couple of years before returning to the United States. During his first stint at the Centre, Marston was a reserved, albeit pleasant figure, who kept mostly to himself.

His rise from postdoc to professor is perhaps one of the most rapid in history, but one thoroughly deserved based upon his stellar research record. His doctoral thesis, for instance, won the prestigious Templer Medal and his 2014 book, *The Indian Army and the End of the Raj,* was runner-up for the prize. Marston returned to the Centre a much more authoritative figure, whose direct approach had the effect of rubbing some up the wrong way, but which played extremely well in the military setting. His forthright manner notwithstanding, few would dispute the fact that he has managed to establish in the MDSP a genuinely world-class graduate program within an extremely short period of time. In particular, Marston's 'Art of War' program, wherein he takes a group of the 'best and brightest' from amongst course members in this program and puts them through a more rigorous and in-depth course of specialised study, has already attracted the attention of some of the world's leading staff colleges and universities.

Two other colleagues took on important leadership roles in the midst of this transition. Ben Schreer, a scholar of German extraction who was appointed to convene the Bachelor of Asia-Pacific Security, became the deputy head of Centre. White had opted not to appoint a deputy during his tenure as head, but it was a position that had traditionally existed within the Centre and that had been occupied by such luminaries as Ball (during O'Neill's time as head) and Horner (during Dibb's tenure). With the Centre expanding rapidly and given my own relative lack of leadership experience, I figured I could use all the help I could get. Schreer's talents as an analyst were quickly recognised around the broader Canberra security community, however, and he understandably took up a good career opportunity at the Australian Strategic Policy Institute (ASPI), lasting little more than 12 months as deputy head of SDSC. Nevertheless, he made an important stabilising contribution during this difficult transition period.

Peter Dean succeeded me as Director of Studies, a post that I had assumed following Ayson's departure in 2009. An historian by training, Dean joined the Centre before I assumed the headship as a replacement for Huisken, who had indicated his intention to retire. Dean came to the Centre from an associate deanship at the University of Notre Dame in Sydney. He was seeking to spend less time on administration and more on research, but his strong administrative pedigree was too tempting for me to resist and he graciously (if probably somewhat reluctantly) accepted the offer to become Director of Studies. Dean has thankfully been able to maintain an impressive trajectory during his time with the Centre, in addition to making an outstanding leadership contribution to the Centre's education programs. During his time as Director of Studies, numbers in the GSSD program rose to record levels and his administrative prowess was recognised elsewhere in the university, as reflected by his appointment as Associate Dean of the College of Asia and the Pacific in June 2015.

The trials of transition notwithstanding, late 2011 was an exciting time as we embarked upon the largest recruitment process in the Centre's history. The first round of job advertisements attracted a field of more than 100 candidates. From this field, we were able to recruit some outstanding new talent. I was especially pleased that we received an application from Amy King, a Rhodes scholar who was completing a doctorate on Sino-Japanese relations at Oxford under the supervision of Rosemary Foot. I'd been following King's work for some time but, evidently, so too had others and I realised that SDSC faced some stiff competition (even from within the ANU itself) were we to succeed in attracting her to the Centre. King had been working closely for a number of years with the hugely respected Peter Drysdale from the Crawford School. Our colleagues in the International Relations department were also recruiting at that time and showed interest in hiring her. Thankfully, the head of International Relations, Bill Tow quickly recognised how keen I was to recruit King and graciously gave me more space than others in the same position might have afforded. Indeed, some of the earlier contributions to this volume have highlighted the often tense historical relationship that has existed between SDSC and International Relations, but relations between our two units have been at their best ever in recent times and that is in large part due to Tow's collegiality and friendship, for which I am immensely grateful.

Another new talent to emerge from the December 2011 recruitment round was Joanne Wallis, an Australian-born, recent University of Cambridge graduate. Wallis is an expert on the South Pacific, an area of obvious strategic importance in which the Centre had not been strong for some time — at least since the days that it counted Greg Fry and David Hegarty amongst its ranks. Wallis had already published in a couple of the leading scholarly journals and went on to convene the Bachelor of Asia-Pacific Security following Schreer's departure — a role that she performed with distinction.

Andrew Carr, a young Australian foreign policy specialist who was competing his doctorate whilst also working at the Lowy Institute for International Policy, primarily on *The Interpreter* blog, also joined the ranks of SDSC out of this initial recruitment round. As noted later in this chapter, amongst his many other contributions Carr has been an immensely valuable addition to the Centre's outreach program, which he convened upon joining SDSC until the end of 2015 when he moved to run the Master's program.

Last, but certainly not least, a relatively long-time associate of the Centre, Jean Bou, who previously worked with Horner on the *Official History of Australian Peacekeeping*, also joined the Centre, where his primary role was to contribute his operational expertise to the newly established MDSP.

Notwithstanding these promising appointments, the Centre lacked scholars working on classical strategic studies, especially in the areas of strategic theory and concepts. Ayson's departure left a gap in this regard and the problem became more acute following Schreer's move to ASPI. With Ball due to retire, this left the Centre with Frühling as the only specialist in this area, despite it being central to SDSC research and teaching. Specialists in this area proved to be in short supply and recruitment was more difficult than anticipated. We were fortunate in the end to find a recently minted doctorate in political science, Charles Miller, who had been working under the supervision of a leading scholar of strategic studies, Peter Feaver, at Duke University.

These appointments made the once top-heavy Centre appear a little light in terms of experience, especially with the looming retirements of a number of SDSC stalwarts such as Ball and Huisken. Joan Beaumont's request for an internal transfer within the university to SDSC thus came at an opportune time. I'd first met Beaumont when she was a visiting fellow with the Centre in the mid-2000s. She subsequently joined ANU as Dean of Education in the College of Arts and Social Sciences, and in that capacity I'd often admired her skill as an academic bureaucrat and enjoyed watching her in action at College Education Committee meetings. But Beaumont was also a well-regarded scholar in her own right who, like Dean, was eager to devote more time to research. I'd encountered Joan most recently as a member of the selection panel for the SDSC headship and it was perhaps in that capacity that her early explorations with White regarding a move to the Centre transpired.

The subsequent recruitment of Russell Glenn, who had spent much of the previous decade working at the RAND Corporation, and the conversion to continuing status of John Blaxland following Ball's retirement added further experience to an otherwise young staff. Even with these important additions, one of the concerns that had weighed most heavily on my mind for much of the past decade — strategists have a naturally anxious disposition, it would seem — was how the Centre was going to survive the transition brought about by the retirement of the genuinely world-class cohort around which it had built much of its national and international reputation. I recall having a discussion on this subject with Ayson during the mid-2000s when we contemplated what the Centre would look like following the eventual retirement from academic life of the likes of Bell, Dibb, Ball, Huisken and Horner. Based upon his career pattern and trajectory prior to joining SDSC, I don't think many expected White to remain at the Centre as long as he has either — and we are immensely fortunate, of course, that he has chosen to do so.

Particularly in the context of this process of generational change, therefore, the recruitment of Evelyn Goh as the Centre's inaugural Shedden Professor in Strategic Policy Studies was critical. I'd known Goh for about a decade prior to her joining the SDSC, having first met at a workshop organised by Tow at Griffith University in the early 2000s, and I'd been a great admirer of her work ever since. A decade on, she was one of the world's leading scholars of Asian security. Her post at SDSC was funded by the Australian Department of Defence, who wanted to give more of a contemporary focus to the professorial position that it had been funding at the Centre for the better part of a decade. Goh fitted the bill perfectly and was the stand-out candidate from the field of applicants for the Shedden Chair.

Goh's arrival added considerable weight to an expanding Centre that was beginning to gather momentum in terms of substantially increasing its research output. Particularly pleasing was the significant increase in the number of books by Centre staff that were being published by highly regarded university presses. Goh has particularly helped to build SDSC's research strength in Asia-Pacific Security, with China a growing area of expertise. This was strengthened in 2016 with the hiring of Bates Gill, formerly of the University of Sydney and a former Director of the Stockholm International Peace Research Institute (SIPRI), and Dr Nina Silove from Stanford University. The Centre's rate of success in applying for major grant funding was also on the rise. Blaxland, for instance, was the only non-American applicant to succeed in winning a prestigious Minerva grant from the US Department of Defense in 2014. Similarly, in collaboration with Gill, Goh was successful in winning a grant as part of the MacArthur Foundation Asia Security Initiative. Beaumont and Wallis each won prestigious Australian Research Council Discovery grants in the 2015 round.

The national and international recognition that is currently being afforded to research produced at the Centre is also notable. The writings of White are particularly worthy of mention in this regard, especially his 2012 book *The China Choice*, which built upon his earlier *Quarterly Essay* 'Power Shift'. White's work is now routinely cited in most of the leading contemporary analyses of Sino-American relations and it has defined the major foreign policy debate in Australia over how Canberra should position itself between a rising China and a United States that is (arguably) in decline. Likewise, Beaumont won the 2014 Prime Minister's Literary Award for Australian History for her highly acclaimed study of Australia during the First World War, *Broken Nation: Australians in the Great War*. Not to be outdone by a fellow historian, David Horner won the 2015 Prime Minister's Literary Award for Australian History for *The Spy Catchers*, which was the first volume in the *Official History of ASIO* series. He also in picked up the St Ermin's Hotel Intelligence Book of the Year for 2015. Yet few achievements could match the inclusion of Ball and White — such different and yet such hugely influential and important Australian strategic thinkers — on the Queen's Birthday Honours List in June

2014. These were extremely special and fitting honours, although it couldn't help but reinforce my sense of inferiority, given that each of my predecessors had now had been awarded the Order of Australia!

The Centre's increasing profile owes a great deal to the outstanding job that Carr has done in leading its outreach program. Two particular areas are worthy of mention. The first is the Centre of Gravity policy papers, a series that has become the Centre's flagship publication and that Carr has been editing since it was launched in November 2013. The Centre previously published hugely successful (and important) SDSC Canberra Papers on Strategy and Defence and Working Papers. Partly for financial reasons and partly due also to the fact that the face of publishing was rapidly changing, White (correctly in my view) opted to abandon these earlier series.

The shorter-form Centre of Gravity series was designed to take their place. Carr and I also toyed with the idea of starting up a blog, but we ultimately felt that we would prefer an intellectually meatier product and that other Australian think tanks (such as ASPI and the Lowy Institute) were already successfully occupying that space. The purpose of the new series was to bring the best national and international minds to bear on the key strategic policy challenges facing Australia. Publication in the series is generally by invitation only and the series has thus far featured a range of world-class contributors that has included Tim Huxley, Robert Ross, Raja Mohan, Rory Medcalf, Geoffrey Till, Dennis Blair, Bob O'Neill, Bates Gill, Michael Green, Bilahari Kausikan, Brad Glosserman and Scott Snyder.

Another feature of SDSC outreach under Carr's leadership was his management of a popular program of public events. These have been held at the ANU and, typically in the case of book launches, at national institutions around Canberra including Parliament House, the Australian War Memorial and the National Museum of Australia. Carr has attracted a range of important and influential figures, including Defence secretaries (Dennis Richardson) and ministers (David Johnston), foreign ministers (Bob Carr), former and future prime ministers (John Howard and Malcolm Turnbull), and other public figures of note (Paul Kelly, Angus Houston, Angus Campbell, John Garnaut and Brendan Nelson, for instance), to speak at these events. He has also been able to attract sizeable audiences. Not since the days of the legendary SDSC conferences of the 1970s and 1980s, which

O'Neill discusses in his chapter, has the Centre been able to boast audience numbers in the several hundreds at its events. Following the launch of the May 2013 Defence White Paper, Carr ran a public lecture featuring Barrie, Richard Brabin-Smith, Dibb and White that attracted over 300 attendees and left standing room only in the Great Hall of University House. Three years later, the Centre repeated the feat with another full house in the Great Hall to hear discussion of the 2016 Defence White Paper.

One of the aspects of the Centre's outreach that concerned me most when I assumed the headship was SDSC's relationship with Defence. Unlike White and Dibb, being a lifelong academic, I had never held any position in government, never mind the role of deputy secretary and author of a Defence white paper. Dibb and White were each generous with sharing their connections. There were those within the department, particularly Michael Shoebridge, the First Assistant Secretary from the Strategic Policy division, who were particularly open to exploring new avenues for engagement with SDSC in those first couple of years. Yet Defence has experienced a series of challenges as a result of internal restructuring and, in my view, its relationship with SDSC thus wasn't anywhere near meeting its full potential. Something seemed to click in early 2015, however, which led to a much greater level of engagement, including the Centre being allowed to fairly regularly deliver in-house roundtables and lectures on site for Defence employees — something that I had been lobbying for at least a couple of years prior. These events benefit significantly from the very evident support given to them by senior officials within Defence, including Tyson Sara, Chris Birrer and Scott Dewar, who attend on a regular basis. They are led largely by Goh in her capacity as Shedden Professor. While requiring a good deal of trust on Defence's part — for which I am both grateful and respectful — they seem to be of benefit to both sides. For this reason, the relationship with Defence is certainly one that I hope continues to deepen in the years ahead.

Notwithstanding all the exciting things that have been happening at the Centre as a result of its increased size, growth did not come without its challenges. One of the consequences of winning the ACSC contract, for instance, was that it led to a situation where the Centre brought on a significant number of military historians due to the heavy operational focus of the MDSP. As Horner notes in his chapter, military history has long been an important element of the Centre's work. Yet, by 2013, following the intense period of recruitment, I noticed that almost half of the Centre's staff was made up of military historians. This was potentially problematic for two reasons. First, it was apparent that colleagues were increasingly speaking about the 'two sides' of the Centre — the historians and the political scientists. While there was no hint of animosity, it wasn't a healthy state of affairs as it set up a dynamic that could pull the Centre in conflicting directions. Secondly, word was getting back to me that some senior officials in the Australian public service were beginning to question the relevance of the Centre relative to other organisations with a more contemporary focus, such as ASPI and the Lowy Institute.

I find the comparison with think tanks puzzling, given that SDSC's mission is to contribute to public and policy debate by offering something different: rigorous, academic research that is accessible and relevant to policymakers. This mission is complementary to that provided by think tankers, who typically react to unfolding developments, often in a pithier manner than that used in academia. Nevertheless, it is a comparison that is often drawn and cannot be ignored.

In an attempt to bring a greater sense of coherence to a rapidly expanding SDSC, in early 2014 I initiated, with the blessing of my colleagues, a small group process to contemplate where the Centre ought optimally to be 5–10 years from now and to develop vision and mission statements for taking us there. I invited Barrie to lead the group, given his experience with running these types of exercises with organisations infinitely larger and more complex than SDSC. I also asked Marston and Goh to join. They are not only the 'next generation' of professors at the Centre, but they also represented well the 'two sides' of the Centre that seemed to be emerging. Bringing these two sides together would thus benefit from their leadership, with the addition of Chris' policy experience of course.

The exercise lasted for the better part of 12 months. We met on a monthly basis, often at national institutions around Canberra. We did quite a bit of preparatory reading in advance of our meetings, both about the history of the Centre and about strategic studies as a field of study. We engaged with each of the living former heads of Centre to seek their counsel and advice, including at least one very productive and enjoyable dinner with O'Neill when he was in town — a dinner that sparked a number of spin-off projects. Most importantly, we reported back regularly to SDSC colleagues at staff meetings and, by December 2014, we were able to gain their agreement to use the following vision and mission to chart the Centre's course for the decade ahead:

Our Mission

The Strategic and Defence Studies Centre (SDSC) is Australia's largest body of scholars dedicated to the analysis of the use of armed force in its political context. As a leading international research institution specialising in strategy and defence, SDSC has a three-part mission:

- To provide 'real world'–focused strategic studies that are research-based, research-led and world-class. Our primary expertise within the broad field of Strategic Studies consists of three related research clusters: Australian defence, military studies, and Asia-Pacific security. Our scholarship in these areas is intended to be recognised internationally and of value to the Australian policy community;

- To prepare and educate the next generation of strategic leaders — military, civilian and academic — in Australia and the Asia-Pacific region by providing world-class graduate and undergraduate programs in strategic and defence studies; and

- To contribute toward a better-informed standard of public debate in Australia and the Asia-Pacific region using high-quality outreach and commentary on issues pertaining to our core areas of expertise.

Our Vision

Founded in 1966, the Strategic and Defence Studies Centre is proud to be counted among the earliest generation of post–World War II research institutions dedicated to the analysis of the use of armed force in its political context. The Centre seeks to build upon our achievements over the past half-century, and to play a leading role in shaping international strategic studies, policies and debates. Within the next decade, SDSC aims to position itself as the leading university-based institution for research, education, and outreach in strategic and defence issues in the Asia-Pacific region. In a fluid and fast-changing strategic milieu, we will invest in the strategic development of our research and teaching programmes in order to play a leading role in defining Strategic Studies for our age. Our goal is to shape the areas of scholarship and policymaking that will inform the leadership of the Asia-Pacific, and equip strategic planners and analysts to be bold and innovative in addressing the challenges of the future.

Some might argue that, framed as such, the mission and vision of the Centre remains too broad. To borrow an analogy from debates regarding the structuring of military forces, a case can be made that the above equates to something akin to a 'balanced force' — one that contains a little bit of everything and thus avoids hard and unpopular choices, which is potentially detrimental to the extent that it spreads effort and resources too thinly.[1]

While there is something to be said for this line of argument, the history both of SDSC and of strategic studies more generally cautions against too narrow a focus. Such a centre, for instance, may not have been sufficiently nimble to respond to the transformational opportunity presented when the ACSC contract came up for tender in 2011. Similarly, a more broadly focused centre may have been better positioned to ride out the ending of the Cold War in the late 1980s and early 1990s. By the same token, a centre focusing only on Australian strategic and defence policy issues would likely not have made the same mark internationally as SDSC has to date, given the limited number of international avenues through which to publish work on issues related to Australia.

Spreading the focus across academic, public and policy debates, as well as across the national, regional and global, will remain a tricky balancing act for a centre such as SDSC. Yet it is one that remains critical to its future viability. The key to SDSC surviving and thriving lies in us remaining a sufficiently broad church within which all members of staff are contributing work of international standard in their chosen area of focus — whether that be publishing in leading, policy-relevant outlets or more traditional academic journals and university presses (or both). This presents challenges of communication as well, and towards that end in May 2016 I restarted the SDSC newsletters that O'Neill in particular had used to successfully inform and promote the Centre's achievements.

The next half-decade will be critical to the future of the Centre and it is my hope that the vision and mission statements we have established will guide us through this period. The end date of the ACSC contract, five years from now, will be a critical turning point.

---

1   For further reading on this point see Hugh White, 'The New Defence White Paper: Why We Need it and What it Needs to Do', *Perspectives*, Apr. 2008.

Ideally, the existing, good working relationship between the ADF and the SDSC will continue and the Centre will remain the ADF partner in the delivery of the MDSP for a long time to come. But sound strategic thinking plans for the worst-case scenario. As such, while continuing to maintain and to build upon the high standards that have been set during the first five years of this program, it is vital that the Centre diversify its income and activities to the point where it could maintain its current size even were the ACSC contract to conclude at the 10-year mark. The Centre's burgeoning Bachelor of International Security Studies program is a promising step towards establishing this certainty. The Centre is also in a position to do much more than it currently does with government and with the private sector, both nationally and internationally, should it so choose. And we must not lose sight of the continued importance of the Master's program as the Centre's most enduring educational offering. The key challenge in any such process of diversification, however, will be to maintain coherence and to continue to adhere to high standards across all areas of activity, especially in our research.

The next five years will also be critical in that the process of staff regeneration will increasingly call upon our early and mid-career SDSC colleagues to fill the big shoes of the past. Promising signs that the Centre, both individually and collectively, is working towards doing that include the work of colleagues such as Blaxland, Dean and Carr to develop strong media profiles, speaking with increasing authority on Australian defence issues. Frühling's appointment to the external expert panel for the 2016 Defence White Paper demonstrates the high regard in which he is held in government and the Australian public service. The Centre does, however, still need to raise its international profile, in particular through publication in the leading academic and policy journals in the field, including *Survival*, *International Security*, *The Washington Quarterly*, *Foreign Affairs*, and the *Journal of Strategic Studies*, to name just a few. In recent decades SDSC colleagues such as Ball, Bell, Dibb and O'Neill appeared frequently in such journals. The onus is now on their SDSC successors to do the same.

Maintaining a strong sense of collegiality will be equally important as the Centre navigates the next few years. While there have been tensions within the Centre over the past 50 years, these have tended to be the exception rather than the rule. Certainly one of the reasons why I have remained at the SDSC for almost 15 years and why the

Centre itself has been so successful is due to the fact that it is such a pleasant, supportive place to work. Such is a rarity in academic circles and is a special characteristic of the Centre that should never be taken for granted.

In the final analysis, the Centre is in a position to thrive over the next decade and beyond. As Dibb's contribution to this volume illustrates all too well, one of the great challenges facing a Centre such as ours is that our fate is tied all too closely to the prevailing strategic environment of the time. SDSC was born, for instance, during the mid-1960s. This period was often referred to as the 'golden age' of strategic studies due to the significant and substantial work being done to develop a more scientific approach to the strategic challenges of the time, particularly the advent and evolution of nuclear weapons. Equally, the end of the Cold War in the late 1980s and early 1990s led to debates over whether strategic studies should even survive. As Dibb notes, this was a dark and dangerous period for the Centre.

By contrast, few would question the relevance of strategy and strategic studies in today's world, where great power competition is on the rise in Asia and where conflict is again breaking out in the Middle East and even in Europe. Some Australian policy elites claim that, as a result, Canberra currently faces its most demanding set of strategic circumstances in our country's relatively short history. As a consequence, a new golden age of strategic studies could well be upon us. As it has done admirably throughout its impressive 50-year history, it is up to SDSC to grab this opportunity with both hands.

50th Anniversary Celebratory Dinner Keynote Speech: 'To See What is Worth Seeing'

Brendan Sargeant, Associate Secretary, Department of Defence

21 July 2016

It is a great privilege and pleasure to be here this evening to give the keynote speech at this celebratory dinner for the 50th Anniversary of the Strategic & Defence Studies Centre.

Let me offer my congratulations on the Centre's 50th anniversary. I think the conference topic, 'New Directions in Strategic Thinking 2.0', is absolutely the right question to be exploring.

It is good to see so many familiar faces here tonight. Some people here I have worked with and they have given me immeasurable help and guidance over the years. Others I know because I have read their work and pondered and learnt from it.

I speak for myself but also on behalf of Defence when I say that the SDSC has made a huge contribution to strategic policy making in Australia over many years, a contribution of incommensurable value to Australia. From my perspective, it has enriched the policy environment and deepened understanding of the world we live in and the nature of the choices that we make as we find our way in that world. Long may it be so.

To speak before such an illustrious audience is a daunting prospect. I am very conscious that almost anything I might talk about is likely to be familiar ground to many in this audience.

I am not going to talk about recent Defence White Papers, or the South China Sea, or the emerging Indo Pacific. If you are looking for advice on government policy, there are plenty of documents available. If you are looking for an expert commentary, there are people in this room better qualified than me.

What I would like to talk about is the importance of strategy and its value to large institutions — and most especially one I know intimately, the Department of Defence.

But first I would like to digress with a couple of anecdotes to set the framework for my discussion.

Many years ago I was haunting a bookshop somewhere in Little Collins Street in Melbourne, a bookshop that no longer exists, and which sold books which were well beyond my price range at that time. I was there one day and I came across a book that had just been published. It was called *The Plains* by Gerald Murnane.

For some reason, I purchased it — spending more money than I could afford — and took it home and read it. It was one of those books that turns you five degrees off centre from the rest of the universe and gives you a completely different picture of the world. Nothing is quite the same after reading it. I think it is one of the great Australian books and it has never left me.

The story is simple enough. A young man who describes himself as a filmmaker decides to leave Outer Australia and journey to a place called Inner Australia.

Inner Australia is the landscape of the plains where a vast and complex culture has been built and sustained by a wealthy landholding aristocracy. These landholders are patrons of the arts and sciences. They are obsessed by the landscape of the plains, which is their landscape. They devote endless resources to discover the true meaning of the plains, to get to an understanding of what they really are, for in knowing the world, their world, they will know themselves. They also know their quest is endless and perhaps futile.

The filmmaker meets these landowners and goes through a process of auditioning. Eventually he is employed by one of the landowners to be a resident filmmaker on his estate. The landowner expects nothing from this filmmaker nut, but believes that he might one day be capable of 'seeing what was worth seeing'. The book then describes the filmmaker's life thereafter. Needless to say, no film is ever made, but the filmmaker goes into an endless and enriching exploration of the plains in this place called Inner Australia.

The book struck me with the force of revelation, for even though it was clearly a work of fantasy or speculative fiction, it described to me absolutely the reality of Australian culture in the world that I was living in. What I realised was that there is a world, but there are many different ways of describing it, and that these can create a richer sense of reality because the process illuminates what may not have been seen. The book is very rich and can be considered in many different ways, but the opposition set up between an Outer Australia, an Australia that is self-satisfied and feels that it knows reality, and an Inner Australia, where the culture is devoted to finding the meaning behind appearances, is worth reflecting on. I will come back to this, but I believe that the work of strategy is the work of this Inner Australia.

My second anecdote relates to my recent visit to Exercise Hamel, a large Army exercise that took place in Cultana, a bleak and beautiful place in South Australia. I visited the exercise and had fun seeing what the Army does when it is being itself.

In the exercise headquarters, the place where the exercise was managed, I saw a map on the wall, which was very familiar to me — it hangs in my office — except that just north-east of the archipelago to our north was another country called Kamaria.

I spent some time contemplating this map, the geography and contours of this imaginary country inserted into a real world, and I remarked to one of my companions that there was an enormous amount of strategic policy history embedded in that simple map. One of the generals said to me: 'They're tough, those Kamarians, we've been fighting them for 40 years.'

What intrigued me, and continues to intrigue me, is how in order to understand ourselves better, we construct an imaginary country against which we define ourselves and test our ideas.

I work in an organisation called the Department of Defence that does many different things every moment of the day. It never sleeps. It never stops. It is relentless. It has its own imperatives and appetites. It has a personality and life independent of those people and organisations that contribute to its being. It is what it is and, in its deepest dreaming, has no desire but to be what it is.

Most of my work is an attempt to help manage this vast enterprise. The practical reality of that is that I make lots of micro decisions or supervise the work of others who also make decisions, or provide advice to more senior decision makers.

In this world strategy can be a distant reality, a quiet voice behind the noise and clutter of the daily routine. Yet I never forget what one of my teachers once said to me: 'Listen to the quiet voice!'

The essence of my management task, as is also that of my colleagues, is to ensure that what this organisation does conforms to government policy and embodies in its activities the strategy that the government has signed up to in its policy documents. These include, most importantly, the White Paper and the subsidiary documents that flow from it, such as the Defence Planning Guidance and the Australian Military Strategy.

What I have seen over the years is a continuing tension between the imperatives of the institution, its personality and its own desires, to speak metaphorically, and the requirements of Government as expressed in policy and strategy. In this sense, strategy is the quiet voice that calls the organisation away from itself and requires that it look out into the world and respond accordingly. For this reason the strategic policy function is central to organisational health and well-being and critical if the organisation is to remain relevant.

One of the features of the current environment is that there is an overwhelming emphasis, and rightly so, on sustaining operations. The challenge is to step back from this immediacy to reflect on the nature and meaning of the larger story that we are telling through what we do. The institution will tell its own story if left to itself. The task of strategy is to make it listen and understand that the reality it sees itself as part of can have many dimensions and actually be something other than what it thinks it is.

We do not have many strategists in Defence. That is not a bad thing, as long as we listen to what they are saying. This I think is the hardest part of working in a large organisation. It is developing the capacity to listen to the other voice — to journey into Inner Australia, so to speak.

What are we doing when we do strategy?

When I reflect on the strategic history of Australia filtered through the development of the Defence organisation, along with the successive documents to chart that development — primarily white papers — what I hear is an ongoing conversation. It is a conversation about who we are. It is a conversation about what sort of country we are, and how we should participate in the world. This conversation takes expression in the capabilities we build, our operational commitments, and in our relationships with other countries.

I could trace the history of strategic thought over the time I have been associated with the Department of Defence. Its main narrative arc goes something like this: in the time after the Vietnam War we started the process of thinking of ourselves as a strategic entity separate from the larger system in which we had participated since Federation. There were many debates, some still alive today. This thinking expressed itself in a policy and a strategy that was called self-reliance and had many dimensions in terms of how we organised the department, began the work of building the modern ADF, and participated in the world.

This was essentially a nationalist project, and an important one. I also think it was part of a larger project of Australia rethinking its place in the world in the post-Vietnam era. The intervention in East Timor might be seen as an expression of that policy and strategy and the arena where its strengths and flaws were highlighted. It is a strategy that has never gone away.

Since Timor and particularly since 9/11, governments have pursued a fairly active engagement of the ADF in many different parts of the world. This reflects, I think, a sense that Australia has global interests and needs to support them, including through the use of the defence force. Our strategy in this context might be seen as a response to globalisation and an attempt to respond to some of the more malignant forces unleashed by globalisation in ways that support our national interests. Whether our strategy has been sufficient for the environment we are in is a debate for another time.

I see the recent White Paper as a culmination of a journey that began decades ago in that it seeks to recognise that Australia is not only a country that lives geographically in the Indo-Pacific, but also has trading and national interests that extend across the world. How we balance the local with the global is an enduring tension in policy-making and strategy development. It is the location of most serious debates about defence policy.

So, I see our strategic thinking as partly the telling of a story about who we are and, more importantly, who we think we are. It is a story that never ends, but will evolve and be reinterpreted over time as events occur and we respond. Most importantly, it is a story we tell both through what we say and what we do.

I am not one of those people who says that the current environment is more difficult or more challenging than the environment faced by our predecessors. I think that is simply being arrogant and historically myopic. Each time has its own demands and every strategic challenge is new to those who have to face it. However, I do believe that we are in one of those moments in history where we are moving from one world to another. The strategic challenge before us is to make the transition successfully.

When you are confronted by a genuinely strategic decision, or there is a genuine strategic change in the environment in which you are in, the challenge is not just a challenge of how you might respond to that environment by taking various forms of action. It is also a challenge to your self-conception, to your sense of who you are, and who you might be. This is why strategic choices are hard and I think difficult for our institutions that can grow comfortable with a sense of things as they are.

It is also why doing the work of strategy is hard. And it should be hard, really hard — emotionally, as well as intellectually.

Many of the contemporary challenges to security are also challenges that go to our sense of what sort of country we are and what we need to become. Some of these challenges have the potential to render the assumptions upon which we take action redundant or meaningless.

To take some examples:

- The assumptions that underpinned the current rules-based global order are increasingly being challenged, and are increasingly challengeable.
- Military power is increasingly a commodity and the ability to generate strategic effects is being democratised. We have all seen what one person with a semi-automatic weapon can do.
- We can not assume that all players in our strategic environment are rational, or share our assumptions about how the world and conflict should be managed.
- It is not so easy anymore to distinguish between the world within our national borders and the world outside.
- We are seeing genuinely transformational technologies — cyber, quantum computing, autonomous systems and so on.

The task of strategy is more complex because it has to speak to many different realities, and many different perceptions of what reality might or should be. It has to do this in a way that helps policy and decision makers thread their way through to a course of action or decision.

Our response to these challenges, along with others that will emerge in coming decades, will change us. How do we understand and manage that change while also responding to what the world brings? Some of our choices will be constrained by our self-conception. We need to understand this as well.

What I worry about is whether we are truly seeing reality. Our institutional imperatives are, in my experience, so potentially powerful that they can blind us to aspects of the world that we live in. Do we prefer to be what we are rather than to consider what we need to be if we are to respond to contemporary realities? What are the costs of the choices that we might need to make and do we really understand what those choices are? In a world of wicked problems, and strategic problems are all wicked, do we prefer our tried and true solution sets rather than seeing what is worth seeing?

When I looked at that map at Exercise Hamel and saw the country of Kamaria, I asked myself a question. In creating an imaginary country that we have used to define and test ourselves against, have we simply created another reflection of what we are and what we are comfortable with being?

When I think of that young man in that imaginary world of the plains commencing his lifelong journey to discover the true meaning of the plains — an impossible but necessary quest — I see it as a wonderful metaphor for the work that all of us do. I am most of all taken by the landowner's implicit request that he come to see what was worth seeing. The landowner understood that this might be the work of a whole lifetime.

Sometimes when I read the work of people who do strategy, including people at the Strategic & Defence Studies Centre, the practical administrator in me gets irritated because it just complicates my decision-making and I prefer a smooth and easy life. It puts in front of me those most terrifying of all questions for an administrator: Have I got it right? Is what we are doing making sense? Do we really understand what we are dealing with?

When I am in a better, less harassed mood, I appreciate how valuable that work is. And I treasure it.

Sometimes I look at much of the writing on strategy and it is like wandering through a library of books about things that have never happened. Sometimes it is quite a strange experience to read these forlorn prophecies that have never come true. Yet, despite this, how important it is that we have these works of imagination, these documents of grim speculation and melancholy advice. They can be books of magic. Sometimes the writing of them ensures that what they talk about does not occur. They intersect with reality to help us understand the reality is more complex and more multidimensional and has more that is imponderable than we are ever quite comfortable with. They help us make choices that change reality.

This conversation, which we call strategy making, helps us understand the world around us and helps us understand the consequences of the choices that we might make or not make. It helps us change the world.

So, to the Strategic and Defence Studies Centre, let us have another 50 years of thinking and conversation and research.

Continue the great work of building a strategic conversation in Australia about who we are and what we might become.

Help us understand the choices and pathways that might take us there.

Help us to see what is worth seeing.

Conference Program – SDSC at 50: New Directions in Strategic Thinking 2.0

The Hall, University House, The Australian National University, Canberra, Australia

Day One — Thursday 21 July 2016

Conference Opening: Dr Brendan Taylor, Head of the Strategic and Defence Studies Centre, and Veronica L. Taylor, Dean of the College of Asia and the Pacific, ANU.

Keynote Address: Strategic Thinking Since 1945: Professor Sir Lawrence Freedman

Student Awards Presentation

Session 1: Strategy and Power

Chair: Professor Michael Wesley

- 21st Century Strategic Order: Dr C. Raja Mohan
- Economics and Strategy: Dr Amy King
- Elements of National Power and Strategic Policy: Major General John J. Frewen
- Great Power Grand Bargains: Myth or Reality?: Professor Evelyn Goh

Session 2: Strategic Thinking: Concepts and Challenges

Chair: Emeritus Professor David Horner

- Old Wine in New Bottles? The Continued Relevance of Cold War Strategic Concepts: Professor Robert Ayson
- Alliances After the Cold War: Professor Thomas Christensen
- Nuclear Strategy After the Cold War: Dr Nicola Leveringhaus

Session 3: Strategy and Domains
Chair: Professor Joan Beaumont
- The Return of Geography: Professor Paul Dibb
- Maritime Strategy in Asia: Dr Euan Graham
- The Evolution of Military Capability in the Indo-Asia-Pacific Region: Dr Tim Huxley

Conference Dinner at the Australian War Memorial

Day Two — Friday 22 July 2016
Welcome: Dr Brendan Taylor

Session 4: Strategic Studies in Practice
Chair: Admiral Chris Barrie
- Strategic Studies in Practice: The Australian Perspective: Professor Hugh White
- Strategic Studies in Practice: The Southeast Asian Perspective: Mr Peter Ho
- Training the Next Generation of Strategic Thinkers: Professor Eliot Cohen

Session 5: New Directions in Strategic Studies
Chair: Professor Daniel Marston
- US Grand Strategy in the Post-Cold War Era: Dr Hal Brands
- The Future of Strategic Studies: Lessons from the Last Golden Age: Professor Sir Hew Strachan
- An Asian School of Strategic Studies?: Professor Amitav Acharya
- The Future of Strategic Studies: The Next Golden Age: Professor Robert O'Neill

Where to from here?: Professor Bates Gill
Concluding Remarks: Dr Brendan Taylor

SDSC 50th Anniversary Public Celebration
- Asia in 2020: War or Peace?: Professor Sir Lawrence Freedman and Professor Eliot Cohen

Appendix: Key SDSC Publications

Occasional books and monographs

Robert O'Neill (ed.), *The Strategic Nuclear Balance: An Australian Perspective* (1974)

H.G. Gelber (ed.), *The Strategic Nuclear Balance 1975: Proceedings of a Conference* (1975)

Peter Hastings (ed.), *New Guinea and Australia: The Pacific and South-East Asia* (1975–1976)

Robert O'Neill (ed.), *The Defence of Australia: Fundamental New Aspects* (1976)

Desmond Ball (ed.), *The Future of Tactical Airpower in the Defence of Australia* (1977)

Jean Fielding and Robert O'Neill, *A Select Bibliography of Australian Military History 1891–1939* (1978)

Jolika Tie, J.O. Langtry and Robert O'Neill, *Australia's Defence Resources: A Compendium of Data* (1978)

David Horner, *Crisis of Command: Australian Generalship and the Japanese threat, 1941–1943* (1978)

J.O. Langtry and Desmond J. Ball, *Controlling Australia's Threat Environment: A Methodology for Planning Australian Defence Force Development* (1979)

Graham Kearns, *Arms for the Poor* (1980)

Ron Huisken, *Defence Resources of South East Asia and the South West Pacific: A Compendium of Data* (1980)

Desmond Ball, *A suitable piece of real estate: American installations in Australia* (1980)

Desmond Ball, *Politics and force levels: The strategic missile program of the Kennedy Administration* (1980)

Ross Babbage, *Rethinking Australia's Defence* (1980)

Robert O'Neill, *Australia in the Korean War 1950–53, Volume I: Strategy and Diplomacy* (1981)

Robert O'Neill and David Horner (eds), *New directions in strategic thinking* (1981)

D.M. Horner, *High Command: Australia's struggle for an independent war strategy, 1939–1945* (1982)

Robert O'Neill and David Horner (eds), *Australian defence policy for the 1980s* (1982)

I.M. Speedy, *Oil and Australia's Security: The Future Fuel Requirements of the Australian Defence Force* (1982)

S.N. Gower, *Options for an Australian Defence Technological Strategy* (1982)

Desmond Ball (ed.), *Strategy & Defence: Australian essays* (1982)

Paul Dibb (ed.), *Australia's External Relations in the 1980s* (1983)

W.S.G. Bateman, *Australia's Overseas Trade Strategic Considerations* (1984)

Desmond Ball, J.O. Langtry and J.D. Stevenson, *Defence Implications of the Alice Springs–Darwin Railway* (1984)

Robert O'Neill, *Australia in the Korean War 1950–53, Volume II: Combat operations* (1985)

W.K. Hancock, *Testimony* (1985)

Andrew Mack, *Peace Research in the 1980s* (1985)

Desmond Ball, J.O. Langtry and J.D. Stevenson, *Defend the North: The case for the Alice Springs–Darwin railway* (1985)

Ernest McNamara, Robin Ward, Desmond Ball, J.O. Langtry and Richard Q. Agnew, *Australia's Defence Resources: A Compendium of Data* (1986)

Paul Dibb, *The Soviet Union: The Incomplete Superpower* (1986)

David Horner, *Australian higher command in the Vietnam War* (1986)

Desmond Ball, *A base for debate: The US satellite station at Nurrungar* (1987)

Desmond Ball and Andrew Mack (eds), *The future of arms control* (1987)

Desmond Ball, *Defence forces and capabilities in the Northern Territory* (1988)

Desmond Ball (ed.), *Air Power: Global developments and Australian perspectives* (1988)

Desmond Ball and Ross Babbage (eds), *Geographic information systems: defence applications* (1989)

Desmond Ball, *The ties that bind: Intelligence cooperation between the UKUSA countries* (1990)

Ross Babbage, *A coast too long: Defending Australia beyond the 1990s* (1990)

David Horner (ed.), *Reshaping the Australian Army: Challenges for the 1990s* (1991)

Desmond Ball (ed.), *Aborigines in the defence of Australia* (1991)

Desmond Ball and Helen Wilson (eds), *Strange neighbours: The Australian-Indonesia relationship* (1991)

David Horner, *General Vasey's War* (1992)

David Horner, *The Gulf commitment: The Australian Defence Force's first war* (1992)

David Horner and Joanna Penglase, *When the war came to Australia: Memories of the Second World War* (1992)

David Horner, *The Commanders: Australian military leadership in the twentieth century* (1992)

David Horner (ed.), *The Battles that shaped Australia:* The Australian's *anniversary essays* (1994)

David Horner (ed.), *Armies and Nation-Building: Past Experience, Future Prospects* (1995)

David Horner, *The Gunners: A history of Australian artillery* (1995)

Paul Dibb and Rhondda Nicholas, *Restructuring the Papua New Guinea Defence Force: Strategic Analysis and Force Structure Principles for a Small State* (1996)

David Horner, *Inside the War Cabinet: Directing Australia's war effort 1939–1945* (1996)

Desmond Ball, and Pauline Kerr, *Presumptive engagement: Australia's Asia-Pacific security policy in the 1990s* (1996)

Desmond Ball (ed.), *The transformation of security in the Asia-Pacific region* (1996)

Desmond Ball, *Burma's Military Secrets: Signals Intelligence (SIGINT) from 1941 to Cyber Warfare* (1998)

Desmond Ball and David Horner, *Breaking the Codes: Australia's KGB network 1944–1950* (1998)

David Horner, *Blamey: The Commander-in-Chief* (1998)

Alan Dupont, *The Environment and Security in Pacific Asia* (1998)

David Horner, *Defence Supremo: Sir Frederick Shedden and the making of Australian defence policy* (2000)

Paul Dibb and Robert D. Blackwill (eds), *America's Asian Alliances* (2000)

Desmond Ball and Hamish McDonald, *Death in Balibo, Lies in Canberra* (2000)

Bob Breen, *Mission Accomplished, East Timor: The Australian Defence Force Participation in the International Forces East Timor* (2001)

David Horner, *Making the Australian Defence Force* (2001)

Bob Breen, *Giving Peace a Chance* (2001)

Alan Dupont, *East Asia Imperilled: Transnational Challenges to Security* (2001)

Desmond Ball and Hamish McDonald, *Masters of Terror: Indonesia's Military and Violence in East Timor in 1999* (2001)

David Horner, *The Second World War (Volume 1): The Pacific* (2002)

David Horner, *SAS: Phantoms of War: A history of the Australian Special Air Service* (2002)

Ron Huisken, *The Road to War on Iraq* (2003)

Coral Bell, *A World Out of Balance: American Ascendancy and International Politics in the 21st Century* (2003)

Ron Huisken, *America and China: A Long Term Challenge for Statesmanship and Diplomacy* (2004)

Ron Huisken, *North Korea: Power Play or Buying Butter with Guns* (2004)

Ron Huisken, *The Threat of Terrorism and Regional Development* (2004)

Edwin Lowe, *Transcending the Cultural Gaps in 21st Century Strategic Planning: The Real Revolution in Military Affairs* (2004)

Clive Williams, *Terrorism Explained: The facts about terrorism and terrorist groups* (2004)

Desmond Ball, *The Boys in Black: The Thahan Phran (Rangers), Thailands Para-military Border Guards* (2004)

Robert Ayson, *Thomas Schelling and the Nuclear Age: Strategy as Social Science* (2004)

Coral Bell, *Living with Giants: Finding Australia's Place in a More Complex World* (2005)

Christopher Michaelsen, *The Use of Depleted Uranium in Operation Iraqi Freedom: A War Crime* (2005)

David Horner, *Strategic Command: General Sir John Wilton and Australia's Asian wars* (2005)

Robert Ayson and Desmond Ball, *Strategy and Security in the Asia-Pacific* (2006)

Desmond Ball and Meredith Thatcher (eds), *A National Asset: Essays commemorating the 40th anniversary of the Strategic and Defence Studies Center* (2006)

Jean Bou, *A Century of Service: A History of the Southport School Army Cadet Unit 1906–2006* (2007)

Ron Huisken and Meredith Thatcher, *History as Policy: Framing the Debate on the Future of Australia's Defence Policy* (2007)

Brendan Taylor (ed.), *Australia as an Asia-Pacific regional power: Friendships in Flux?* (2007)

Coral Bell and Meredith Thatcher, *Rembembering Hedley* (2008)

Jean Bou and David Horner, *Duty First: A History of the Royal Australian Regiment* (2008)

Jean Bou, Peter Dennis, Jeffrey Grey, Robin Prior and Ewan Morris, *The Oxford Companion to Australian Military History* (2008)

Daniel Marston and Carter Malkasian, *Counterinsurgency in Modern Warfare* (2008)

Desmond Ball, Gary Waters and Ian Dudgeon, *Australia and Cyber-Warfare* (2008)

Stephan Frühling, *A History of Australian Strategic Policy Since 1945* (2009)

David Worher with Neil Thomas, *In action with the SAS* (2009)

David Horner, Peter Londey and Jean Bou (eds), *Australian Peacekeeping: Sixty years in the field* (2009)

David Horner, *Australia's Military History for Dummies* (2010)

Brendan Taylor, *American Sanctions in the Asia-Pacific* (2010)

Brendan Taylor, *Sanctions as Grand Strategy* (2010)

David Connery, *Crisis Policymaking: Australia and the East Timor Crisis of 1999* (2010)

Ron Huisken, *Introducing China: The World's oldest great power charts its next comeback* (2010)

Desmond Ball and Kwa Chong Guan, *Assessing Track 2 Diplomacy in the Asia-Pacific Region* (2010)

Cameron Crouch, *Managing Terrorism and Insurgency: Regeneration, recruitment and attrition* (2010)

Hugh White, *Power Shift: Australia's Future between Washington and Beijing* (2010)

Jean Bou, *Light Horse: A History of Australia's Mounted Arm* (2010)

David Horner, *Australia and the 'New World Order': From peacekeeping to peace enforcement: 1988–1991* (2011)

Stephan Frühling, Michael Clarke and Andrew O'Neill, *Australia's Uranium Trade: The Domestic and Foreign Policy Challenges of a Contentious Export* (2011)

Brendan Taylor, Nicholas Farelly and Sheryn Lee, *Insurgent Intellectual: Essays in Honour of Professor Desmond Ball* (2012)

David Brewster, *India as an Asia Pacific Power* (2012)

Peter J. Dean, *Australia 1942: In the Shadow of War* (2012)

Jean Bou, *MacArthur's Secret Bureau: The story of the Central Bureau, General MacArthur's signals intelligence organisation* (2012)

Hugh White, *The China Choice: Why America Should Share Power* (2012)

Joan Beaumont, *Broken Nation Australians in the Great War* (2013)

Leszek Buszynski, *Negotiating with North Korea: The Six Party Talks and the Nuclear Issue* (2013)

Peter J. Dean, *Australia 1943: The Liberation of New Guinea* (2013)

Evelyn Goh, *The Struggle for Order: Hegemony, Hierarchy, and Transition in Post-Cold War East Asia* (2013)

Desmond Ball, *Tor Chor Dor: Thailand's Border Patrol Police (BPP), Volume 1: History, Organisation, Equipment, and Personnel* (2013)

Desmond Ball, *Tor Chor Dor: Thailand's Border Patrol Police (BPP), Volume 2: Activities and Prospects* (2013)

Desmond Ball and Keiko Tamura, *Breaking Japanese Diplomatic Codes: David Sissons and D Special Section during the Second World War* (2013)

Rhys Crawley, *Climax at Gallipoli: The Failure of the August Offensive* (2014)

Joanne Wallis, *Constitution Making during State Building* (2014)

Stephan Frühling, *Defence Planning and Uncertainty: Preparing for the next Asia-Pacific war* (2014)

David Brewster, *India's Ocean: The Story of India's Bid for Regional Leadership* (2014)

John Blaxland, *The Australian Army from Whitlam to Howard* (2014)

Daniel Marston, *The Indian Army and the End of the Raj* (2014)

David Horner, *The Spy Catchers: The Official History of ASIO, 1949–1963, Volume I* (2014)

David Horner and John Connor, *'The Good International Citizen': Australian Peacekeeping in Asia, Africa and Europe 1991–1993* (2014)

Desmond Ball and Sheryn Lee (eds), *Power and International Relations: Essays in Honour of Coral Bell* (2014)

Peter J. Dean, Stephan Frühling and Brendan Taylor (eds), *Australia's Defence: Towards an New Era?* (2014)

Andrew Carr, *Winning the Peace: Australia's Campaign to Change the Asia-Pacific* (2015)

Russell Glenn, *Rethinking Western Approaches to Counterinsurgency: Lessons from post-colonial conflict* (2015)

Desmond Ball and Richard Tanter, *The Tools of Owatatsumi: Japan's Ocean Surveillance and Coastal Defence Capabilities* (2015)

Mark Johnston, *Stretcher-Bearers: Saving Australians from Gallipoli to Kokoda* (2015)

Stephan Frühling, Andrew O'Neil and Michael Clark, *Australia's Nuclear Policy: Reconciling Strategic, Economic and Normative Interests* (2015)

John Lee, *Trends in Southeast Asia: Strategic Possibilities and Limitations for Abe's Japan in Southeast Asia* (2015)

J. Beaumont, L. Grant and A. Pegram (eds), *Beyond Surrender: Australian Prisoners of War in the Twentieth Century* (2015)

Peter J. Dean (ed.), *Australia 1944–45: Victory in the Pacific* (2015)

John Blaxland, *The Protest Years: The Official History of ASIO 1963–1975, Volume II* (2015)

John Blaxland (ed.), *East Timor Intervention: A Retrospective on INTERFET* (2015)

Desmond Ball and Sheryn Lee (eds), *Geography, Power, Strategy and Defence Policy: Essays in Honour of Paul Dibb* (2016)

Daniel Marston and Tamara Leahy (eds), *War, Strategy and History: Essays in Honour of Professor Robert O'Neill* (2016)

John Blaxland and Rhys Crawley, *The Secret Cold War: The Official History of ASIO 1975–1989, Volume III* (2016)

Amy King, *China-Japan Relations after World War Two: Empire, Industry and War 1949–1971* (2016)

Joanne Wallis and Andrew Carr (eds), *Asia-Pacific Security: An Introduction* (2016)

Evelyn Goh (ed.), *Rising China's Influence in Developing Asia* (2016)

Jean Bou (ed.), *The AIF in Battle: How the Australian Imperial Force Fought* (2016)

Jean Bou and Peter Dennis, *The Australian Imperial Force: Volume 5 The Centenary History of Australia and the Great War* (2016)

Russell W. Glenn, *Short War in a Perpetual Conflict: Implications of Israeli's 2014 Operation Protective Edge for the Australian Army* (2016)

Peter J. Dean, Stephan Frühling and Brendan Taylor (eds), *Australia's American Alliance: Towards an New Era?* (2016)

Centre of Gravity Papers

COG 1 Rory Medcalf, *Pivoting the Map: Australia's Indo-Pacific System* (2012)

COG 2 Tim Huxley, *Australia's Defence Engagement with Southeast Asia* (2012)

COG 3 Paul Dibb, *A Sovereign Submarine Capability in Australia's Grand Strategy* (2012)

COG 4 Hugh White, *An Australia–Japan Alliance?* (2012)

COG 5 Robert S. Ross, *The US Pivot to Asia and the Implications for Australia* (2013)

COG 6 Amy King, Andrew Carr, Joanne Wallis, Emma Campbell, John Blaxland and Peter J. Dean, *A New Flank: Fresh Perspectives For the Next Defence White Paper* (2013)

COG 7 C. Raja Mohan, *The Sino-Indian Border Dispute and Asian Security* (2013)

COG 8 Dennis C. Blair, *What Should We Do about China?* (2013)

COG 9 Robert O'Neill, *Preparing to Face Our Next Enemy* (2013)

COG 10 Bates Gill, *Alliances under Austerity: What does America Want?* (2013)

COG 11 Evelyn Goh, *Ringing in a New Order? Hegemony, Hierachy and Transition in East Asia* (2013)

COG 12 Richard Brabin-Smith, *Contingencies and Warning Time* (2013)

COG 13 David Kang, *North Korea: Beyond the Nuclear Challenge* (2013)

COG 14 Geoffrey Till, *Outgoing Australia?* (2014)

COG 15 Andrew Carr and Peter J. Dean, *What the Next Defence White Paper Should Do about the Budget* (2014)

COG 16 Paul Dibb, *The Geopolitical Implications of Russia's Invasion of the Ukraine* (2014)

COG 17 Brendan Taylor, John Blaxland, Hugh White, Nick Bisley, Peter Leahy and See Seng Tan, *Defence Diplomacy: Is the Game Worth the Candle?* (2014)

COG 18 Robert Ayson and Desmond Ball, *Escalation in Northeast Asia: A Strategic Challenge for Australia* (2014)

COG 19 Micah Zenko, *The A Word: An Accomodationist Strategy for US–China Relations* (2015)

COG 20 Evelyn Goh, Greg Fealy and Ristian Atriandi Supriyanto, *A Strategy Towards Indonesia* (2015)

COG 21 Brad Glosserman and Scott A. Snyder, *Unlocking the Japan– ROK Relationship: The Key is National Identity* (2015)

COG 22 Bilahari Kausiakan, *Asia's Strategic Challenge: Manoeuvring Between the US and China* (2015)

COG 23 Michael J. Green, Peter J. Dean, Brendan Taylor and Zack Cooper, *The ANZUS Alliance in an Ascending Asia* (2015)

COG 24 Peter Varghese, *Australia and the challenges of weapons of mass destruction* (2016)

COG 25 Anthea McCarthy-Jones, *Mexican drug cartels and dark networks: An emerging threat to Australia's national security* (2016)

COG 26 John Blaxland, *MANIS: Time for a new forum to sweeten regional cooperation* (2016)

COG 27 Youngshik Bong, Jaehyon Lee, Shafiah F. Muhibat, Christine Susanna Tjhin, Brendan Taylor and William T. Tow, *The South China Sea: Middle Power Perspectives* (2016)

Canberra Papers on Strategy and Defence

CP1 Alex Hunter, *Oil Supply in Australia's Defence Strategy* (1968)

CP2 Geoffrey Jukes, *The Strategic Situation in the 1980s* (1968)

CP3 J.L. Richardson, *Australia and the Non-proliferation Treaty* (1968)

CP4 Ian Bellany and J.L. Richardson, *An Australian Nuclear Force* (1969)

CP5 P.H. Partridge, *Educating for the Profession of Arms* (1969)

CP6 Robert J. O'Neill, *The Strategy of General Giap Since 1964* (1969)

CP7 T.B. Millar, *Soviet Policies in the Indian Ocean Area* (1970)

CP8 Ian Bellany and J.L. Richardson, *Australian Defence Procurement* (1970)

CP9 John Welfield, *Japan and Nuclear China* (1970)

CP10 Robert J. O'Neill, *The Army in Papua New Guinea* (1971)

CP11 Darcy McGaurr, *Conscription and Australian Military Capability* (1971)

CP12 Peter King, *The Strategy of Total Withholding* (1971)

CP13 W.A.C. Adie, *Chinese Military Thinking under Mao Tse-tung* (1972)

CP14 Geoffrey Jukes, *The Development of Soviet Strategic Thinking Since 1945* (1972)

CP15 Hedley Bull, *The Moscow Agreements and Strategic Arms Limitation* (1973)

CP16 Ron Huisken, *Arms Limitation in South-East Asia: A Proposal* (1977)

CP17 Ross Babbage, Desmond Ball, J.O. Langtry and Robert O'Neill, *The Development of Australian Army Officers for the 1980s* (1978)

CP18 Mohammed Ayoob, *The Horn of Africa: Regional Conflict and Super Power Involvement* (1978)

CP19 Shelton Kodikara, *Strategic Factors in Interstate Relations in South Asia* (1979)

CP20 Ron Huisken, *The Cruise Missile and Arms Control* (1980)

CP21 K.R. Singh, *The Persian Gulf: Arms and Arms Control* (1980)

CP22 S.D. Muni, *Arms Build-up and Development: Linkages in the Third World* (1980)

CP23 Pervaiz Iqbal Cheema, *Conflict and Cooperation in the Indian Ocean* (1980)

CP24 John J. Weltman, *Nuclear Weapons Spread and Australian Policy* (1981)

CP25 Talukder Maniruzzaman, *The Security of Small States in the Third World* (1982)

CP26 Ram Rajan Subramanian, *Nuclear Proliferation in South Asia: Security in the 1980s* (1982)

CP27 Geoffrey Hartnell, *The Problem of Command in the Australian Defence Force Environment* (1983)

CP28 Ngok Lee, *The Chinese People's Liberation Army 1980–82: Modernisation, Strategy and Politics* (1983)

CP29 F.A. Mediansky and Dianne Court, *The Soviet Union in Southeast Asia* (1984)

CP30 K.K. Nair, *ASEAN — Indochina Relations Since 1975: The Politics of Accommodation* (1984)

CP31 K.V. Kesavan, *Japanese Defence Policy Since 1976: Latest Trends* (1984)

CP32 Neville G. Brown, *Limited World War?* (1984)

CP33 D.B. Nichols, *The Strategic Implications for Australia of the New Law of the Sea* (1985)

CP34 Tony Godfrey-Smith, *Low Level Conflict Contingencies and Australian Defence Policy* (1985)

CP35 Andrew Selth, *The Terrorist Threat to Diplomacy: An Australian Perspective* (1986)

CP36 Ray Sunderland, *Problems in Australian Defence Planning* (1986)

CP37 Robert D. Glasser, *Nuclear Pre-emption and Crisis Stability 1985–1990* (1986)

CP38 Michael Ward, *The Regional Concentration of Defence Spending: Issues, Implications and Policies Concerning Defence Infrastructure Development in Australia* (1986)

CP39 Russell Solomon, *The Role of Japan in United States Strategic Policy for Northeast Asia* (1986)

CP40 D.M. Horner, *Australian Higher Command in the Vietnam War* (1986)

CP41 F.W. Speed, *Command Structure of the Australian Defence Force* (1987)

CP42 Amin Saikal, *The Afghanistan Conflict: Gorbachev's Options* (1987)

CP43 Desmond Ball, *Australia's Secret Space Programs* (1988)

CP44 Cathy Downes, *High Personnel Turnover: The Australian Defence Force is not a Limited Liability Company* (1988)

CP45 Ross Babbage, *Should Australia Plan to Defend Christmas and Cocos Islands?* (1988)

CP46 Desmond Ball (ed.), *US Bases in the Philippines: Issues and Implications* (1988)

CP47 Desmond Ball, *Soviet Signals Intelligence (SIGINT)* (1989)

CP48 D.M. FitzGerald, *The Vietnam People's Army: Regularization of Command 1975–1988* (1989)

CP49 Desmond Ball, *Australia and the Global Strategic Balance* (1989)

CP50 J.C. Blaxland, *Organising an Army: The Australian Experience 1957–1965* (1989)

CP51 Richard A. Higgott, *The Evolving World Economy: Some Alternative Security Questions for Australia* (1989)

CP52 Peter Donovan, *Defending the Northern Gateway* (1989)

CP53 Desmond Ball, *Soviet Signals Intelligence (SIGINT): Intercepting Satellite Communications* (1989)

CP54 Gary Brown, *Breaking the American Alliance: An Independent National Security Policy for Australia* (1989)

CP55 Cathy Downes, *Senior Officer Professional Development in the Australian Defence Force: Constant Study to Prepare* (1989)

CP56 Desmond Ball, *Code 777: Australia and the US Defense Satellite Communications System (DSCS)* (1989)

CP57 Gary Klintworth (ed.), *China's Crisis: The International Implications* (1989)

CP58 Gary Brown, *Index to Parliamentary Questions on Defence* (1989)

CP59 W.A.G. Dovers, *Controlling Civil Maritime Activities in a Defence Contingency* (1989)

CP60 David Hegarty and Peter Polomka (eds), *The Security of Oceania in the 1990s Vol. 1: Views from the Region* (1989)

CP61 Ross Babbage, *The Strategic Significance of Torres Strait* (1990)

CP62 P.J. Criss and D.J. Schubert, *The Leading Edge: Air Power in Australia's Unique Environment* (1990)

CP63 Desmond Ball and J.O. Langtry (eds), *The Northern Territory in the Defence of Australia: Geography, History, Economy, Infrastructure, and Defence Presence* (1990)

CP64 Gary Klintworth (ed.), *Vietnam's Withdrawal From Cambodia: Regional Issues and Realignments* (1990)

CP65 Ken Ross, *Prospects for Crisis Prediction: A South Pacific Case Study* (1990)

CP66 Peter Polomka (ed.), *Bougainville: Perspectives on a Crisis* (1990)

CP67 F.N. Bennett, *The Amateur Managers: A Study of the Management of Weapons System Projects* (1990)

CP68 Peter Polomka (ed.), *The Security of Oceania in the 1990s Vol. 2: Managing Change* (1990)

CP69 Desmond Ball (ed.), *Australia and the World: Prologue and Prospects* (1990)

CP70 Bilveer Singh, *Singapore's Defence Industries* (1990)

CP71 Gary Waters (ed.), *RAAF Air Power Doctrine: A Collection of Contemporary Essays* (1990)

CP72 Stephen Henningham and Desmond Ball (eds), *South Pacific Security: Issues and Perspectives* (1991)

CP73 J.O. Langtry and Desmond Ball (eds), *The Northern Territory in the Defence of Australia: Strategic and Operational Considerations* (1991)

CP74 Gary Waters, *The Architect of Victory: Air Campaigns for Australia* (1991)

CP75 Gary Klintworth (ed.), *Modern Taiwan in the 1990s* (1991)

CP76 Desmond Ball and Helen Wilson (eds), *New Technology: Implications for Regional and Australian Security* (1991)

CP77 David Horner (ed.), *Reshaping the Australian Army: Challenges for the 1990s* (1991)

CP78 Desmond Ball, *The Intelligence War in the Gulf* (1991)

CP79 Desmond Ball, *Provocative Plans: A Critique of US Strategy for Maritime Conflict in the North Pacific* (1991)

CP80 Desmond Ball, *Soviet SIGINT: Hawaii Operation* (1991)

CP81 Owen Wilkes, Megan van Frank and Peter Hayes, *Chasing Gravity's Rainbow: Kwajalein and US Ballistic Missile Testing* (1991)

CP82 Alan Dupont, *Australia's Threat Perceptions: A Search for Security* (1991)

CP83 Desmond Ball, *Building Blocks for Regional Security: An Australian Perspective on Confidence and Security Building Measures (CSBMs) in the Asia/Pacific Region* (1991)

CP84 Alan Dupont, *Australia's Security Interests in Northeast Asia* (1991)

CP85 Paul Lee, *Finance and Financial Policy in Defence Contingencies* (1991)

CP86 Alan Hinge, *Mine Warfare in Australia's First Line of Defence* (1992)

CP87 Peter J. Rimmer, *Hong Kong's Future as a Regional Transport Hub* (1992)

CP88 Paul Dibb, *The Conceptual Basis of Australia's Defence Planning and Force Structure Development* (1992)

CP89 Desmond Ball and David Horner (eds), *Strategic Studies in a Changing World: Global, Regional and Australian Perspectives* (1992)

CP90 J. Mohan Malik, *The Gulf War: Australia's Role and Asia-Pacific Responses* (1992)

CP91 Desmond Ball, *Defence Aspects of Australia's Space Activities* (1992)

CP92 Philip Methven, *The Five Power Defence Arrangements and Military Cooperation among the ASEAN States: Incompatible Models for Security in Southeast Asia?* (1992)

CP93 T.M. Boyce, *Infrastructure and Security: Problems of Development in the West Sepik Province of Papua New Guinea* (1992)

CP94 Desmond Ball and Helen Wilson (eds), *Australia and Space* (1992)

CP95 David W. Beveridge, *LANDFORCE 2010: Some Implications of Technology for ADF Future Land Force Doctrine, Leadership and Structures* (1992)

CP96 Wayne Gobert, *The Origins of Australian Diplomatic Intelligence in Asia, 1933–1941* (1992)

CP97 Peter Polomka, *Japan as Peacekeeper: Samurai State or New Civilian Power?* (1992)

CP98 Coral Bell, *The Post-Soviet World: Geopolitics and Crises* (1992)

CP99 Bob Lowry, *Indonesian Defence Policy and the Indonesian Armed Forces* (1993)

CP100 Ken Ross, *Regional Security in the South Pacific: The Quarter-century 1970–95* (1993)

CP101 R.J. May, *The Changing Role of the Military in Papua New Guinea* (1993)

CP102 Sam Bateman and Dick Sherwood (eds), *Strategic Change and Naval Roles: Issues for a Medium Naval Power* (1993)

CP103 J.N. Mak, *ASEAN Defence Reorientation 1975–1992: The Dynamics of Modernisation and Structural* Change (1993)

CP104 Coral Bell (ed.), *The United Nations and Crisis Management: Six Studies* (1994)

CP105 Dick Sherwood (ed.), *Operational and Technological Developments in Maritime Warfare: Implications for the Western Pacific* (1994)

CP106 Nicola Baker, *More than Little Heroes: Australian Army Air Liaison Officers in the Second World War* (1994)

CP107 Matthew Gubb, *Vanuatu's 1980 Santo Rebellion: International Responses to a Microstate Security Crisis* (1994)

CP108 M.C.J. Welburn, *The Development of Australian Army Doctrine 1945–1964* (1994)

CP109 Dick Sherwood, *The Navy and National Security: The Peacetime Dimension* (1994)

CP110 Desmond Ball, *Signals Intelligence (SIGINT) in South Korea* (1995)

CP111 Sandy Gordon and Stephen Henningham (eds), *India Looks East: An Emerging Power and Its Asia-Pacific Neighbours* (1995)

CP112 Coral Bell (ed.), *Nation, Region and Context: Studies in Peace and War in Honour of Professor T.B. Millar* (1995)

CP113 Andrew Selth, *Transforming the Tatmadaw: The Burmese Armed Forces since 1988* (1996)

CP114 Sam Bateman and Stephen Bates (eds), *Calming the Waters: Initiatives for Asia Pacific Maritime Cooperation* (1996)

CP115 Ken Anderson and Paul Dibb, *Strategic Guidelines for Enabling Research and Development to Support Australian Defence* (1996)

CP116 Sandy Gordon, *Security and Security Building in the Indian Ocean Region* (1996)

CP117 Desmond Ball, *Signals Intelligence (SIGINT) in South Asia: India, Pakistan, Sri Lanka (Ceylon)* (1996)

CP118 Sam Bateman and Stephen Bates (eds), *The Seas Unite: Maritime Cooperation in the Asia Pacific Region* (1996)

CP119 David Stevens (ed.), *In Search of a Maritime Strategy: The Maritime Element in Australian Defence Planning since 1901* (1997)

CP120 Helen Hookey and Denny Roy (eds), *Australian Defence Planning: Five Views from Policy Makers* (1997)

CP121 Kim Richard Nossal and Carolynn Vivian, *A Brief Madness: Australia and the Resumption of French Nuclear Testing* (1997)

CP122 Greg Austin (ed.), *Missile Diplomacy and Taiwan's Future: Innovations in Politics and Military Power* (1997)

CP123 Peter Chalk, *Grey-Area Phenomena in Southeast Asia: Piracy, Drug Trafficking and Political Terrorism* (1997)

CP124 Sam Bateman and Stephen Bates (eds), *Regional Maritime Management and Security* (1998)

CP125 Alan Dupont (ed.), *The Environment and Security: What are the Linkages?* (1998)

CP126 R.N. Bushby, *'Educating an Army': Australian Army Doctrinal Development and the Operational Experience in South Vietnam, 1965–72* (1998)

CP127 Greg Mills, *South Africa and Security Building in the Indian Ocean Rim* (1998)

CP128 Maree Reid, *The Shape of Things to Come: The US–Japan Security Relationship in the New Era* (1998)

CP129 Sam Bateman and Stephen Bates (eds), *Shipping and Regional Security* (1998)

CP130 Karl Claxton, *Bougainville 1988–98: Five Searches for Security in the North Solomons Province of Papua New Guinea* (1998)

CP131 Desmond Ball and Amitav Acharya (eds), *The Next Stage: Preventive Diplomacy and Security Cooperation in the Asia-Pacific Region* (1999)

CP132 Sam Bateman (ed.), *Maritime Cooperation in the Asia-Pacific Region: Current Situation and Prospects* (1999)

CP133 Desmond Ball (ed.), *Maintaining the Strategic Edge: The Defence of Australia in 2015* (1999)

CP134 R.W. Cable, *An Independent Command: Command and Control of the 1st Australian Task Force in Vietnam* (2000)

CP135 Andrew Tan, *Armed Rebellion in the ASEAN States: Persistence and Implications* (2000)

CP136 Andrew Selth, *Burma's Secret Military Partners* (2000)

CP137 Peter McLennan, *Where Are They When You Need Them? Support Arrangements for Deployed Air Power* (2001)

CP138 Bilveer Singh, *ASEAN, the Southeast Asia Nuclear-Weapon-Free Zone and the Challenge of Denuclearisation in Southeast Asia: Problems and Prospects* (2001)

CP139 Desmond Ball, *The Council for Security Cooperation in the Asia Pacific (CSCAP): Its Record and Its Prospects* (2001)

CP140 John Hutcheson, *Wars of Conscience. Human Rights, National Security and Australia's Defence Policy* (2001)

CP141 Bilveer Singh, *Civil–military Relations in Democratising Indonesia: The Potentials and Limits to Change* (2001)

CP142 Bob Breen, *Giving Peace a Chance: Operation Lagoon, Bougainville, 1994: A Case of Military Action and Diplomacy* (2002)

CP143 Nick Jans with David Schmidtchen, *The Real C-Cubed Culture, Careers and Climate and How They Affect Capability* (2002)

CP144 David Dickens (ed.), *The Human Face of Security: Asia-Pacific Perspectives* (2002)

CP145 Hamish McDonald, *Desmond Ball, James Dunn, Gerry van Klinken, David Bourchier and Richard Tanter, Masters of Terror: Indonesia's Military and Violence in East Timor in 1999* (2002)

CP146 David Capie, *Small Arms Production and Transfers in Southeast Asia* (2002)

CP147 Clive Williams and Brendan Taylor (eds), *Countering Terror: New Directions Post September 11* (2003)

CP148 Ron Huisken, *The Road to War on Iraq* (2003)

CP149 Brek Batley, *The Complexities of Dealing with Radical Islam in Southeast Asia: A Case Study of Jemaah Islamiyah* (2003)

CP150 Andrew Selth, *Burma's Muslims: Terrorists or Terrorised?* (2003)

CP151 Stephan Frühling, *Ballistic Missile Defence for Australia: Policies, Requirements and Options* (2003)

CP152 Bilveer Singh, *ASEAN, Australia and the Management of the Jemaah Islamiyah Threat* (2003)

CP153 Bilveer Singh, *Arming the Singapore Armed Forces (SAF): Trends and Implications* (2003)

CP154 Andrew Selth, *Burma's North Korean Gambit: A Challenge to Regional Security?* (2004)

CP155 Edwin H. Lowe, *Transcending the Cultural Gaps in 21st Century Strategic Analysis and Planning: The Real Revolution in Military Affairs* (2004)

CP156 Christian Enemark, *Disease Security in Northeast Asia: Biological Weapons and Natural Plagues* (2004)

CP157 Reuben R.E. Bowd, *A Basis for Victory: The Allied Geographical Section, 1942–1946* (2005)

CP158 Matt Weiner, *An Afghan 'Narco-State'? Dynamics, Assessment and Security Implications of the Afghan Opium Industry* (2005)

CP159 Gary Waters and Desmond Ball, *Transforming the Australian Defence Force (ADF) for Information Superiority* (2005)

CP160 Blair Tidey, *Forewarned Forearmed: Australian Specialist Intelligence Support in South Vietnam, 1966–1971* (2006)

CP161 Paul Dibb, *Essays on Australian Defence* (2006)

CP162 Maryanne Kelton, *New Depths in Australia–US Relations: The Collins Class Submarine Project* (2006)

CP163 Christian Enemark (ed.), *Ethics of War in a Time of Terror* (2006)

CP164 Brendan Taylor, *Anthony Milner and Desmond Ball, Track 2 Diplomacy in Asia: Australian and New Zealand Engagement* (2006)

SDSC Working Papers

WP1 Robert O'Neill, *The Defence of Continental Australia* (1978)

WP2 J.O. Langtry, *Manpower Alternatives for the Defence Forces* (1978)

WP3 Robert O'Neill, *Structural Changes for a More Self-reliant National Defence* (1978)

WP4 Desmond J. Ball, *Australia and Nuclear Non-proliferation* (1978)

WP5 Desmond J. Ball, *American Bases: Some Implications for Australian Security* (1978)

WP6 T.B. Millar, *The Political–Military Relationship in Australia* (1979)

WP7 Mohammed Ayoob, *The Two Faces of Political Islam: Pakistan and Iran Compared* (1979)

WP8 Ron Huisken, *Cost-effectiveness and the B–1 Strategic Bomber* (1979)

WP9 Philip Towle, *Limiting the Use of Conventional Weapons: Prospects for the 1979 U.N. Conference (Future of Incendiaries, Cluster Bombs, High Vvelocity Rifles, Fuel-air Explosives and Land Mines)* (1979)

WP10 Robert O'Neill, *The Structure of Australia's Defence Force* (1979)

WP11 Michael McGwire, *Australia as a Regional Seapower: An External View* (1979)

WP12 Mohammed Ayoob, *The Indian Ocean Littoral: Projections for the 1980s* (1979)

WP13 Desmond J. Ball, *The Australian Tactical Fighter Force: Prologue and Prospects* (1979)

WP14 Philip Towle, *Non-aligned Criticisms of Western Security Policies* (1979)

WP15 Milton Osborne, *Aggression and Annexation: Kampuchea's Condemnation of Vietnam* (1979)

WP16 Mohammed Ayoob, *Blueprint for a Catastrophe: Conducting Oil Diplomacy by 'Other Means' in the Middle East and the Persian Gulf* (1979)

WP17 Desmond J. Ball, *Developments in US Strategic Nuclear Policy Under the Carter Administration* (1979)

WP18 Philip Towle, *Australian Policy in the Committee on Disarmament* (1980)

WP19 Pervaiz Iqbal Cheema, *Pakistan's Quest for Nuclear Technology* (1980)

WP20 Philip Towle, *The Strategy of War by Proxy* (1980)

WP21 Mohammed Ayoob, *The Politics of Resurgent Islam* (1980)

WP22 J.O. Langtry, *The Status of Australia's Defence Preparedness* (1980)

WP23 Philip Towle, *Arms Control and Detente* (1980)

WP24 Robert O'Neill, *Australia's Future Defence Requirements* (1980)

WP25 Gordon Lawrie, *Problems of Flexible Response* (1980)

WP26 Ravindra Tomar, *Development of the Indian Navy: An Overstated Case?* (1980)

WP27 T.B. Millar, *Global and Regional Changes and Their Defence Implications for Australia to the Year 2000* (1980)

WP28 D.M. Horner, *Australia and Allied Intelligence in the Pacific in the Second World War* (1980)

WP29 Robert O'Neill, *The Strategic Environment in the 1980s* (1980)

WP30 Robert O'Neill, *Australia's Strategic Options in the 1980s* (1980)

WP31 Desmond J. Ball, *The Future of the Strategic Balance* (1981)

WP32 Ram Rajan Subramanian, *South Asia of the 1980s: Implications of Nuclear Proliferation* (1981)

WP33 Mohammed Ahsen Chaudhri, *Pakistan's Security in a Changing World* (1981)

WP34 Ray Sunderland, *Australia's Next War?* (1981)

WP35 Mohammed Ayoob, *Defusing the Middle East Time Bomb: A State for the Palestinians* (1981)

WP36 Desmond Ball, *US Installations in Australia* (1981)

WP37 Mohammed Ayoob, *Southwest Asia: Beginnings of a New Cold War* (1981)

WP38 D.B. Nicholls, *The Visiting Force Acts: A Study in Inter-Service Command and Discipline* (1981)

WP39 J.O. Langtry, *Australia's Civil Defence in Perspective* (1981)

WP40 Robert O'Neill, *Strategic Studies and Political Scientists: Strategic Studies and its Critics Re-visited* (1981)

WP41 Donald H. McMillen, *China and the 'Contending Barbarians': Beijing's View of the Contemporary World Order* (1981)

WP42 T.B. Millar, *The Role of Academics in Defence and Foreign Policy* (1981)

WP43 Desmond O'Connor, *Problems of Research and Development Relating to the Defence of Northern Australia* (1981)

WP44 Desmond O'Connor, *The Future of Defence Science and Technology in Australia: General Considerations* (1981)

WP45 Paul Dibb, *Soviet Capabilities, Interests and Strategies in East Asia in the 1980s* (1982)

WP46 Sreedhar, *Flashpoints in the Gulf* (1982)

WP47 Sreedhar, *Security Profile of the Gulf* (1982)

WP48 Sreedhar, *The Gulf Oil Scene* (1982)

WP49 Sreedhar, *Arms Flow into the Gulf: Process of Buying Security* (1982)

WP50 Donald H. McMillen, *The Urumqui Military Region: Defence and Security in China's West* (1982)

WP51 Donald H. McMillen, *China's Political Battlefront: Deng Xiaoping and the Military* (1982)

WP52 Desmond O'Connor, *Technological Forecasting in the Australian Military Environment* (1982)

WP53 Ean Higgins, *Options and Constraints for US Far Eastern Policy: Five Issue Areas* (1982)

WP54 Bilveer Singh, *The Development of Moscow–Hanoi Relations Since the Vietnam War: The View from Singapore* (1982)

WP55 Samuel Makinda, *Kenya's Role in the Somali–Ethiopian Conflict* (1982)

WP56 H.G. Gelber, *Australia, the U.S., and the Strategic Balance: Some Comments on the Joint Facilities* (1982)

WP57 Gary Brown and Derek Woolner, *New Aircraft Carrier for the Royal Australian Navy?* (1982)

WP58 Desmond Ball, *Issues in Strategic Nuclear Targeting: Target Selection and Rates of Fire* (1982)

WP59 Alan Robertson, *The Need for an Australian Aircraft Carrier Capability* (1982)

WP60 T.B. Millar, *The State of the Western Alliance* (1982)

WP61 T.B. Millar, *Controlling the Spread of Nuclear Weapons* (1982)

WP62 John J. Weltman, *Managing Nuclear Polarity* (1982)

WP63 J.O. Langtry, *Aspects of Leadership in a Modern Army* (1983)

WP64 Iqbal Singh, *Indian Ocean: A Zone of Peace or Power Play?* (1983)

WP65 Paul Dibb, *World Political and Strategic Trends over the Next 20 Years—Their Relevance to Australia* (1983)

WP66 J.O. Langtry and Desmond Ball, *The Concept of Force Multipliers and the Development of the Australian Defence Force* (1983)

WP67 Tim Huxley, *Indochina and Insurgency in the ASEAN States, 1975–1981* (1983)

WP68 Warwick J. Graco, *Problems and Prospects in Managing Servicemen's Careers: A Review* (1983)

WP69 Warwick J. Graco, *Performance-Based Training: An Explanation and Reappraisal* (1983)

WP70 J.O. Langtry, *The Civil Infrastructure in the Defence of Australia: A Regional Approach* (1983)

WP71 V.J. Kronenberg and Hugh Smith, *Civil–Military Relations in Australia; The Case of Officer Education, 1965–1980* (1983)

WP72 Donald Hugh McMillen, *China in Asian International Relations* (1983)

WP73 T.B. Millar, *The Resolution of Conflict and the Study of Peace* (1983)

WP74 Major General K.J. Taylor, *The Australian Army of Today and Tomorrow* (1983)

WP75 Greg Fry, *A Nuclear-free Zone for the Southwest Pacific: Prospects and Significance* (1984)

WP76 Zakaria Haji Ahmad, *War and Conflict Studies in Malaysia: The State of the Art* (1984)

WP77 Derek Woolner, *Funding Australia's Defence* (1984)

WP78 Ray Sunderland, *Australia's Changing Threat Perceptions* (1984)

WP79 I.F. Andrew, *Human Resources and Australian Defence* (1984)

WP80 Ray Sunderland, *Australia's Emerging Regional Defence Strategy* (1984)

WP81 Paul Dibb, *The Soviet Union as a Pacific Military Power* (1984)

WP82 Samuel M. Makinda, *Soviet Policy in the Red Sea Region* (1984)

WP83 Andrew Mack, *The Political Economy of Global Decline: America in the 1980s* (1984)

WP84 Andrew Selth, *Australia and the Republic of Korea: Still Allies or Just Good Friends?* (1984)

WP85 F.W. Speed, *Command in Operations of the Australian Defence Force* (1984)

WP86 F.W. Speed, *Australian Defence Force Functional Commands* (1984)

WP87 Harry Gelber, *Mr Reagan's 'Star Wars': Towards a Strategic Era?* (1984)

WP88 Tim Huxley, *The ASEAN States' Defence Policies, 1975–81: Military Responses to Indochina?* (1984)

WP89 Geoffrey Jukes (trans.), *The Civil Defence of the USSR: This Everybody Must Know and Understand A Handbook for the Population* (1984)

WP90 Paul Dibb, *Soviet Strategy Towards Australia, New Zealand and Oceania* (1984)

WP91 Andrew Selth, *Terrorist Studies and the Threat to Diplomacy* (1985)

WP92 Andrew Selth, *Australia and the Terrorist Threat to Diplomacy* (1985)

WP93 Peter J. Murphy, Civilian Defence: *A Useful Component of Australia's Defence Structure?* (1985)

WP94 Ray Sunderland, *Australia's Defence Forces—Ready or Not?* (1985)

WP95 Ray Sunderland, *Selecting Long-Term Force Structure Objectives* (1985)

WP96 W.H. Talberg, *Aspects of Defence: Why Defence?* (1985)

WP97 F.W. Speed, *Operational Command by the Chief of the Defence Force* (1985)

WP98 Ron Huisken, *Deterrence, Strategic Defence and Arms Control* (1985)

WP99 Desmond Ball, *Strategic Defenses: Concepts and Programs* (1986)

WP100 Stanley S. Schaetzel, *Local Development of Defence Hardware in Australia* (1986)

WP101 Air Marshal S.D. Evans, *Air Operations in Northern Australia* (1986)

WP102 Andrew Selth, *International Terrorism and Australian Foreign Policy: A Survey* (1986)

WP103 Andrew MacIntyre, *Internal Aspects of Security in Asia and the Pacific: An Australian Perspective* (1986)

WP104 B.C. Brett, *Rethinking Deterrence and Arms Control* (1986)

WP105 J.A.C. Mackie, *Low-level Military Incursions: Lessons of the Indonesia–Malaysia 'Confrontation' Episode, 1963–66* (1986)

WP106 Paul Keal, *Japan's Role in United States Strategy in the Pacific* (1986)

WP107 Gary Brown, *Detection of Nuclear Weapons and the US Non-disclosure Policy* (1986)

WP108 Ross Babbage, *Managing Australia's Contingency Spectrum for Defence Planning* (1986)

WP109 Ross Babbage, *Australia's Approach to the United States Strategic Defense Initiative (SDI)* (1986)

WP110 Ross Babbage, *Looking Beyond the Dibb Report* (1986)

WP111 Gary Klintworth, *Mr Gorbachev's China Diplomacy* (1986)

WP112 Samina Yasmeen, *The Comprehensive Test Ban Treaty: Verification Problems* (1986)

WP113 Ross Babbage, *The Future of the Australian–New Zealand Defence Relationship* (1986)

WP114 Gary Klintworth, *Kim Il Sung's North Korea: At the Crossroads* (1986)

WP115 Gary Brown, *The Australian Defence Force in Industrial Action Situations: Joint Service Plan 'CABRIOLE'* (1986)

WP116 Hugh Smith, *Conscientious Objection to Particular Wars: The Australian Approach* (1986)

WP117 Gary Klintworth, *Vietnam's Withdrawal from Cambodia* (1987)

WP118 Harry G. Gelber, *Nuclear Arms Control After Reykjavik* (1987)

WP119 Harry G. Gelber, *A Programme for the Development of Senior Officers of the Australian Defence Force* (1987)

WP120 Ciaran O'Faircheallaigh, *The Northern Territory Economy: Growth and Structure 1965–1985* (1987)

WP121 Robert A. Hall, *Aborigines and Torres Strait Islanders in the Second World War* (1987)

WP122 Tim Huxley, *The ASEAN States' Internal Security Expenditure* (1987)

WP123 J.O. Langtry, *The Status of Australian Mobilization Planning in 1987* (1987)

WP124 Gary Klintworth, *China's India War: A Question of Confidence* (1987)

WP125 Samina Yasmeen, *India and Pakistan: Why the Latest Exercise in Brinkmanship?* (1987)

WP126 David Hegarty, *Small State Security in the South Pacific* (1987)

WP127 David Hegarty, *Libya and the South Pacific* (1987)

WP128 Ross Babbage, *The Dilemmas of Papua New Guinea (PNG) Contingencies in Australian Defence Planning* (1987)

WP129 Ross Babbage, *Christmas and the Cocos Islands: Defence Liabilities or Assets?* (1987)

WP130 Amitav Acharya, *The Gulf War and 'Irangate': American Dilemmas* (1987)

WP131 J.O. Langtry, *The Defence Para-military Manpower Dilemma: Militia or Constabulary?* (1987)

WP132 J.O. Langtry, *'Garrisoning' the Northern Territory: The Army's Role* (1987)

WP133 J.O. Langtry, *The Case for a Joint Force Regional Command Headquarters in Darwin* (1987)

WP134 Desmond Ball, *The Use of the Soviet Embassy in Canberra for Signals Intelligence (SIGINT) Collection* (1987)

WP135 Desmond Ball and J.O. Langtry, *Army Manoeuvre and Exercise Areas in the Top End* (1987)

WP136 Graeme Neate, *Legal Aspects of Defence Operations on Aboriginal Land in the Northern Territory* (1987)

WP137 Gary Brown, *The ANZUS Alliance—The Case Against* (1987)

WP138 Desmond Ball, *Controlling Theater Nuclear War* (1987)

WP139 J.O. Langtry, *The Northern Territory in the Defence of Australia: Geostrategic Imperatives* (1987)

WP140 J.O. Langtry, *The Ambient Environment of the Northern Territory: Implications for the Conduct of Military Operations* (1987)

WP141 Samina Yasmeen, *Is the Non-aligned Movement Really Non-aligned?* (1987)

WP142 A.D. Garrisson, *The Australian Submarine Project: An Introduction to Some General Issues* (1987)

WP143 Commander Stephen Youll, *The Northern Territory in the Defence of Australia: Naval Considerations* (1987)

WP144 J.O. Langtry, *The Northern Territory in the Defence of Australia: A Potential Adversary's Perceptions* (1987)

WP145 Samuel Makinda, *The INF Treaty and Soviet Arms Control* (1987)

WP146 J.O. Langtry, *Infrastructure Development in the North: Civil–Military Interaction* (1987)

WP147 David Hegarty, *South Pacific Security Issues: An Australian Perspective* (1987)

WP148 Brice Pacey, *The Potential Role of Net Assessment in Australian Defence Planning* (1988)

WP149 Ian Wilson, *Political Reform and the 13th Congress of the Communist Party of China* (1988)

WP150 Andrew Mack, *Australia's Defence Revolution* (1988)

WP151 R.H. Mathams, *The Intelligence Analyst's Notebook* (1988)

WP152 Thomas-Durell Young, *Assessing the 1987 Australian Defence White Paper in the Light of Domestic Political and Allied Influences on the Objective of Defence Self-reliance* (1988)

WP153 Clive Williams, *The Strategic Defense Initiative (SDI): The North Pacific Dimension* (1988)

WP154 W.A.G. Dovers, *Australia's Maritime Activities and Vulnerabilities* (1988)

WP155 Ross Babbage, *Coastal Surveillance and Protection: Current Problems and Options for the Future* (1988)

WP156 Warwick J. Graco, *Military Competence: An Individual Perspective* (1988)

WP157 Desmond Ball, *Defence Forces and Capabilities in the Northern Territory* (1988)

WP158 Ross Babbage, *The Future of United States Maritime Strategy in the Pacific* (1988)

WP159 David Hodgkinson, *Inadvertent Nuclear War: The US Maritime Strategy and the 'Cult of the Offensive'* (1988)

WP160 Viberto Selochan, *Could the Military Govern the Philippines?* (1988)

WP161 Tas Maketu, *Defence in Papua New Guinea: Introductory Issues* (1988)

WP162 Deborah Wade-Marshall, *The Northern Territory in the Defence of Australia: Settlement History, Administration and Infrastructure* (1988)

WP163 Thomas-Durell Young, *The Diplomatic and Security Implications of ANZUS Naval Relations, 1951–1985* (1988)

WP164 Matthew Gubb, *How Valid was the Criticism of Paul Dibb's 'Review of Australia's Defence Capabilities'?* (1988)

WP165 Leszek Buszynski, *ASEAN: Security Issues of the 1990s* (1988)

WP166 Tim Huxley, *Brunei's Defence Policy and Military Expenditure* (1988)

WP167 Warwick J. Graco, *Manpower Considerations in Mobilizing the Australian Army for Operational Service* (1988)

WP168 John Chappell, *The Geographic Context for Defence of the Northern Territory* (1988)

WP169 Cathy Downes, *Social, Economic and Political Influences Upon the Australian Army of the 1990s* (1988)

WP170 Robert Ayson, *Activities of the Soviet Fishing Fleet: Implications for Australia* (1988)

WP171 Matthew Gubb, *The Australian Military Response to the Fiji Coup: An Assessment* (1988)

WP172 Malcolm Mackintosh, *Gorbachev and the Soviet Military* (1988)

WP173 Malcolm Mackintosh, *Gorbachev's First Three Years* (1988)

WP174 Jim Sanday, *South Pacific Culture and Politics: Notes on Current Issues* (1988)

WP175 Brigadier P.J. Greville, *Why Australia Should Not Ratify the New Law of War* (1989)

WP176 Peter Donovan, *The Northern Territory and the Defence of Australia: Historical Overview* (1989)

WP177 David Hegarty, *Papua New Guinea: At the Political Crossroads?* (1989)

WP178 Gary Klintworth, *China's Indochina Policy* (1989)

WP179 Gary Klintworth and Ross Babbage, *Peacekeeping in Cambodia: An Australian Role?* (1989)

WP180 David W. Beveridge, *Towards 2010: Security in the Asia-Pacific, an Australian Regional Strategy* (1989)

WP181 Gary Klintworth, *The Vietnamese Achievement in Kampuchea* (1989)

WP182 Leszek Buszynski, *The Concept of Political Regulation in Soviet Foreign Policy: The Case of the Kampuchean Issue* (1989)

WP183 A.C. Kevin, *Major Power Influences on the Southeast Asian Region: An Australian View* (1989)

WP184 Denis McLean and Desmond Ball, *The ANZAC Ships* (1989)

WP185 David Hegarty, *Stability and Turbulence in South Pacific Politics* (1989)

WP186 Stephen J. Cimbala, *Nuclear War Termination: Concepts, Controversies and Conclusions* (1989)

WP187 Peter Jennings, *Exercise Golden Fleece and the New Zealand Military: Lessons and Limitations* (1989)

WP188 Desmond Ball, *Soviet Signals Intelligence (SIGINT): Listening to ASEAN* (1989)

WP189 Thomas-Durell Young, *ANZUS: Requiescat in Pace?* (1989)

WP190 Harry G. Gelber, *China's New Economic and Strategic Uncertainties; and the Security Prospects* (1989)

WP191 David Hegarty and Martin O'Hare, *Defending the Torres Strait: The Likely Reactions of Papua New Guinea and Indonesia to Australia's Initiatives* (1989)

WP192 Commodore H.J. Donohue, *Maritime Lessons from the 1971 Indo-Pakistan War* (1989)

WP193 Ross Babbage, *The Changing Maritime Equation in the Northwest Pacific* (1989)

WP194 Ross Babbage, *More Troops for our Taxes? Examining Defence Personnel Options for Australia* (1989)

WP195 You Ji and Ian Wilson, *Leadership Politics in the Chinese Party– Army State: The Fall of Zhao Ziyang* (1989)

WP196 Jan Prawitz, *The Neither Confirming Nor Denying Controversy* (1989)

WP197 Stanley S. Schaetzel, *The Death of an Aircraft: The A–10 Debacle* (1989)

WP198 Stanley S. Schaetzel, *Fourteen Steps to Decision — or, the Operations of the Defence Department* (1989)

WP199 Stanley S. Schaetzel, *The Coastal Exposure of Australia* (1989)

WP200 Stanley S. Schaetzel, *The Space Age and Australia* (1989)

WP201 Jim Sanday, *The Military in Fiji: Historical Development and Future Role* (1989)

WP202 Stephanie Lawson, *The Prospects for a Third Military Coup in Fiji* (1989)

WP203 John C. Dorrance, *Strategic Cooperation and Competition in the Pacific Islands: An American Assessment* (1990)

WP204 John C. Dorrance, *The Australian–American Alliance Today: An American Assessment of the Strategic/Security, Political and Economic Dimensions* (1990)

WP205 John C. Jeremy, *Naval Shipbuilding: Some Australian Experience* (1990)

WP206 Alan Dupont, *Australia and the Concept of National Security* (1990)

WP207 John C. Dorrance, *The Soviet Union and the Pacific Islands: An American Assessment and Proposed Western Strategy* (1990)

WP208 Stephen Bates, *Security Perceptions in the South Pacific: Questionnaire Results* (1990)

WP209 Alan Henderson, *SLCMs, Naval Nuclear Arms Control and US Naval Strategy* (1990)

WP210 Gary Klintworth, *Cambodia and Peacekeeping: 1990* (1990)

WP211 B.G. Roberts, *Economic Life Analysis of Defence Systems and Equipment* (1990)

WP212 Ian MacFarling, *Military Aspects of the West New Guinea Dispute, 1958–1962* (1990)

WP213 A.C. Kevin, *Southeast Asia Beyond the Cambodia Settlement: Sources of Political and Economic Tensions and Conflict, Trends in Defence Spending, and Options for Cooperative Engagement* (1990)

WP214 Norman MacQueen, *The South Pacific: Regional Subsystem or Geographical Expression?* (1990)

WP215 Norman MacQueen, *United Nations Peacekeeping in a Transforming System* (1990)

WP216 Gary Klintworth, *Iraq: International Law Aspects* (1990)

WP217 Gary Klintworth, *Vietnam's Strategic Outlook* (1990)

WP218 Andrew Selth, *'Assisting the Defence of Australia': Australian Defence Contacts with Burma, 1945–1987* (1990)

WP219 Peter Edwards, *Australia and the Crises in Laos, 1959–61* (1990)

WP220 J.O. Langtry, *The Northern Territory in the Defence of Australia: The Civil–Military Nexus* (1990)

WP221 You Ji, *Jiang Zemin's Leadership and Chinese Elite Politics after 4 June 1990* (1990)

WP222 You Xu and You Ji, *In Search of Blue Water Power: The PLA Navy's Maritime Strategy in the 1990s and Beyond* (1990)

WP223 Kusuma Snitwongse, *Southeast Asia Beyond a Cambodia Settlement: Conflict or Cooperation?* (1990)

WP224 Andrew Selth, *Politically Motivated Violence in the Southwest Pacific* (1990)

WP225 Sandy Gordon, *India's Strategic Posture: 'Look East' or 'Look West'?* (1991)

WP226 Gary Brown, *Index to Parliamentary Questions on Defence for the Period 1989 to 1990* (1991)

WP227 Katherine Bullock, *Australia and Papua New Guinea: Foreign and Defence Relations Since 1975* (1991)

WP228 J.O. Langtry, *The Wrigley Report: An Exercise in Mobilisation Planning* (1991)

WP229 Desmond Ball, *Air Power, the Defence of Australia and Regional Security* (1991)

WP230 Desmond Ball, *Current Strategic Developments and Implications for the Aerospace Industry* (1991)

WP231 Gary Klintworth, *Arms Control and Great Power Interests in the Korean Peninsula* (1991)

WP232 Ian Wilson, *Power, the Gun and Foreign Policy in China since the Tiananmen Incident* (1991)

WP233 Amin Saikal and Ralph King, *The Gulf Crisis: Testing a New World Order?* (1991)

WP234 Desmond Ball and Commodore Sam Bateman, *An Australian Perspective on Maritime CSBMs in the Asia-Pacific Region* (1991)

WP235 Andrew Selth, *Insurgency and the Transnational Flow of Information: A Case Study* (1991)

WP236 Sandy Gordon, *India's Security Policy: Desire and Necessity in a Changing World* (1991)

WP237 Lieutenant Colonel T.M. Boyce, *The Introduction of the Civilian National Service Scheme for Youth in Papua New Guinea* (1991)

WP238 Shaun Gregory, *Command, Control, Communications and Intelligence in the Gulf War* (1991)

WP239 Stephen R. Heder, *Reflections on Cambodian Political History: Backgrounder to Recent Developments* (1991)

WP240 Gary Klintworth, *The Asia-Pacific: More Security, Less Uncertainty, New Opportunities* (1991)

WP241 Matthew L. James, *A History of Australia's Space Involvement* (1991)

WP242 John Wells, *Antarctic Resources: A Dichotomy of Interest* (1991)

WP243 Gary Klintworth, *'The Right to Intervene' in the Domestic Affairs of States* (1991)

WP244 Greg Johannes, *An Isolated Debating Society: Australia in Southeast Asia and the South Pacific* (1992)

WP245 Di Hua, *Recent Developments in China's Domestic and Foreign Affairs: The Political and Strategic Implications for Northeast Asia* (1992)

WP246 E.A. Olsen, *The Evolution of US Maritime Power in the Pacific* (1992)

WP247 Gary Brown, *Index to Parliamentary Questions on Defence, 1991* (1992)

WP248 Elizabeth Ward, *Call Out the Troops: An Examination of the Legal Basis for Australian Defence Force Involvement in 'Non-Defence' Matters* (1992)

WP249 Charles E. Heller, *The Australian Defence Force and the Total Force Policy* (1992)

WP250 James Wood, *Mobilisation: The Gulf War in Retrospect* (1992)

WP251 James Wood, *Mobilisation: The Benefits of Experience* (1992)

WP252 Andrew Butfoy, *Strategic Studies and Extended Deterrence in Europe: A Retrospective* (1992)

WP253 Ken Granger, *Geographic Information and Remote Sensing Technologies in the Defence of Australia* (1992)

WP254 Andrew Butfoy, *The Military Dimension of Common Security* (1992)

WP255 Gary Klintworth, *Taiwan's New Role in the Asia-Pacific Region* (1992)

WP256 Paul Dibb, *Focusing the CSBM Agenda in the Asia/Pacific Region: Some Aspects of Defence Confidence Building* (1992)

WP257 Stewart Woodman, *Defence and Industry: A Strategic Perspective* (1992)

WP258 Leszek Buszynski, *Russia and the Asia-Pacific Region* (1992)

WP259 Bruce Vaughn, *National Security and Defence Policy Formation and Decision-Making in India* (1992)

WP260 Stewart Woodman, *A Question of Priorities: Australian and New Zealand Security Planning in the 1990s* (1992)

WP261 Peter I. Peipul, *Papua New Guinea-Australia Defence and Security Relations* (1992)

WP262 Paul Dibb, *The Regional Security Outlook: An Australian Viewpoint* (1992)

WP263 Liu Jinkun, *Pakistan's Security Concerns: A Chinese Perspective* (1992)

WP264 Andrew Mack and Desmond Ball, *The Military Build-up in the Asia-Pacific Region: Scope, Causes and Implications for Security* (1992)

WP265 W.S.G. Bateman and R.J. Sherwood, *Principles of Australian Maritime Operations* (1992)

WP266 Gary Klintworth, *Sino-Russian Detente and the Regional Implications* (1992)

WP267 Peter Jennings, *Australia and Asia-Pacific Regional Security* (1992)

WP268 Gary Klintworth, *Cambodia's Past, Present and Future* (1993)

WP269 Wing Commander R.W. Grey, *Australia's Aerial Surveillance Programme in the South Pacific: A Review and New Options* (1993)

WP270 Desmond Ball, *Strategic Culture in the Asia-Pacific Region (With Some Implications for Regional Security Cooperation)* (1993)

WP271 Stewart Woodman, *Australian Security Planning at the Crossroads: The Challenge of the Nineties* (1993)

WP272 Gary Brown, *Index to Parliamentary Questions on Defence, 1992* (1993)

WP273 Desmond Ball, *Trends in Military Acquisitions in the Asia-Pacific Region: Implications for Security and Prospects for Constraints and Controls* (1993)

WP274 Wing Commander R.W. Grey, *A Proposal for Cooperation in Maritime Security in Southeast Asia* (1993)

WP275 Captain Russ Swinnerton, *The Preparation and Management of Australian Contingents in UN Peacekeeping Operations* (1993)

WP276 Paul Dibb, *The Future of Australia's Defence Relationship with the United States* (1993)

WP277 Geoffrey Jukes, *Russia's Military and the Northern Territories Issue* (1993)

WP278 Captain Russ Swinnerton and Desmond Ball, *A Regional Regime for Maritime Surveillance, Safety and Information Exchanges* (1993)

WP279 Tim Huxley, *The Political Role of the Singapore Armed Forces' Officer Corps: Towards a Military-Administrative State?* (1993)

WP280 Desmond Ball, *The East Coast Armaments Complex (ECAC) Location Project: Strategic and Defence Aspects* (1994)

WP281 Captain Russ Swinnerton, *Rules of Engagement in Maritime Operations* (1994)

WP282 Paul Dibb, *The Political and Strategic Outlook, 1994–2003: Global, Regional and Australian Perspectives* (1994)

WP283 Gary Brown, *Index to Parliamentary Questions on Defence, 1993* (1994)

WP284 Nobuyuki Takaki, *New Dimensions to the Japan–Australia Relationship: From Economic Preference to Political Cooperation* (1994)

WP285 Sandy Gordon, *Winners and Losers: South Asia After the Cold War* (1995)

WP286 Jim Rolfe, *Australia and New Zealand: Towards a More Effective Defence Relationship* (1995)

WP287 Sheng Lijun, *China's Policy Towards the Spratly Islands in the 1990s* (1995)

WP288 Paul Dibb, *How to Begin Implementing Specific Trust-Building Measures in the Asia-Pacific Region* (1995)

WP289 Andrew Selth, *Burma's Arms Procurement Programme* (1995)

WP290 Desmond Ball, *Developments in Signals Intelligence and Electronic Warfare in Southeast Asia* (1995)

WP291 D.N. Christie, *India's Naval Strategy and the Role of the Andaman and Nicobar Islands* (1995)

WP292 Naoko Sajima, *Japan and Australia: A New Security Partnership?* (1996)

WP293 Brigadier Chris Roberts, *Chinese Strategy and the Spratly Islands Dispute* (1996)

WP294 John McFarlane and Karen McLennan, *Transnational Crime: The New Security Paradigm* (1996)

WP295 Desmond Ball, *Signals Intelligence (SIGINT) in North Korea* (1996)

WP296 Paul Dibb, *The Emerging Geopolitics of the Asia-Pacific Region* (1996)

WP297 Jack McCaffrie, *Maritime Strategy into the Twenty-First Century: Issues for Regional Navies* (1996)

WP298 Coral Bell, *The Cold War in Retrospect: Diplomacy, Strategy and Regional Impact* (1996)

WP299 Bob Lowry, *Australia-Indonesia Security Cooperation: For Better or Worse?* (1996)

WP300 John Chipman, *Reflections on American Foreign Policy Strategy* (1996)

WP301 Jaana Karhilo, *New Requirements for Multilateral Conflict Management by UN and Other Forces: Nordic Responses* (1996)

WP302 Bill Houston, *Developing Army Doctrine in the Post-Cold War Era* (1996)

WP303 Desmond Ball, *The Joint Patrol Vessel (JPV): A Regional Concept for Regional Cooperation* (1996)

WP304 Harry G. Gelber, *Australian-American Relations after the Collapse of Communism* (1996)

WP305 Yukio Satoh, *Policy Coordination for Asia-Pacific Security and Stability* (1996)

WP306 Paul Dibb, *Force Modernisation in Asia: Towards 2000 and Beyond* (1997)

WP307 Jörn Dosch, *PMC, ARF and CSCAP: Foundations for a Security Architecture in the Asia-Pacific?* (1997)

WP308 Andrew Selth, *Burma's Intelligence Apparatus* (1997)

WP309 Andrew Selth, *Burma's Defence Expenditure and Arms Industries* (1997)

WP310 Adam Cobb, *Australia's Vulnerability to Information Attack: Towards a National Information Policy* (1997)

WP311 Desmond Ball, *Australia, the US Alliance, and Multilateralism in Southeast Asia* (1997)

WP312 Yukio Satoh, *From Distant Countries to Partners: The Japan–Australia Relationship* (1997)

WP313 Andrew Selth, *The Burma Navy* (1997)

WP314 Andrew Tan, *Problems and Issues in Malaysia–Singapore Relations* (1997)

WP315 Andrew Selth, *The Burma Air Force* (1997)

WP316 Brigadier Mike Smith, *Australia's National Security into the Twenty-First Century* (1997)

WP317 Paul Dibb, Alliances, *Alignments and the Global Order: The Outlook for the Asia-Pacific Region in the Next Quarter-Century* (1997)

WP318 Greg Mills, *The South African National Defence Force: Between Downsizing and New Capabilities?* (1998)

WP319 Sheng Lijun, *The Evolution of China's Perception of Taiwan* (1998)

WP320 Jim Sanday, *UN Peacekeeping, UNIFIL and the Fijian Experience* (1998)

WP321 Alan Dupont, *The Future of the ASEAN Regional Forum: An Australian View* (1998)

WP322 Andrew Tan, *Singapore's Defence Policy in the New Millennium* (1998)

WP323 Alexei Mouraviev, *Responses to NATO's Eastward Expansion by the Russian Federation* (1998)

WP324 Paul Dibb, *The Remaking of Asia's Geopolitics* (1998)

WP325 Desmond Ball and Mohan Malik, *The Nuclear Crisis in Asia: The Indian and Pakistani Nuclear Programmes* (1998)

WP326 Pauline Kerr, *Researching Security in East Asia: From 'Strategic Culture' to 'Security Culture'* (1998)

WP327 Maung Aung Myoe, *Building the Tatmadaw: The Organisational Development of the Armed Forces in Myanmar, 1948–98* (1998)

WP328 Alan Dupont, *Drugs, Transnational Crime and Security in East Asia* (1998)

WP329 Paul Dibb, *The Relevance of the Knowledge Edge* (1998)

WP330 Desmond Ball, *The US–Australian Alliance: History and Prospects* (1999)

WP331 Desmond Ball, *Implications of the East Asian Economic Recession for Regional Security Cooperation* (1999)

WP332 Daniel T. Kuehl, *Strategic Information Warfare: A Concept* (1999)

WP333 Desmond Ball, *Security Developments and Prospects for Cooperation in the Asia-Pacific Region, with Particular Reference to the Mekong River Basin* (1999)

WP334 Andrew Selth, *Burma and Weapons of Mass Destruction* (1999)

WP335 John McFarlane, *Transnational Crime and Illegal Immigration in the Asia-Pacific Region: Background, Prospects and Countermeasures* (1999)

WP336 Desmond Ball, *Burma and Drugs: The Regime's Complicity in the Global Drug Trade* (1999)

WP337 Paul Dibb, *Defence Strategy in the Contemporary Era* (1999)

WP338 Andrew Selth, *The Burmese Armed Forces Next Century: Continuity or Change?* (1999)

WP339 Maung Aung Myoe, *Military Doctrine and Strategy in Myanmar: A Historical Perspective* (1999)

WP340 Desmond Ball, *The Evolving Security Architecture in the Asia-Pacific Region* (1999)

WP341 John McFarlane, *The Asian Financial Crisis: Corruption, Cronyism and Organised Crime* (1999)

WP342 Maung Aung Myoe, *The Tatmadaw in Myanmar since 1988: An Interim Assessment* (1999)

WP343 Tony Kevin, *Cambodia and Southeast Asia* (1999)

WP344 Herman Kraft, *The Principle of Non-Intervention and ASEAN: Evolution and Emerging Challenges* (2000)

WP345 Paul Dibb, *Will America's Alliances in the Asia-Pacific Region Endure?* (2000)

WP346 Maung Aung Myoe, *Officer Education and Leadership Training in the Tatmadaw: A Survey* (2000)

WP347 Paul Dibb, *The Prospects for Southeast Asia's Security* (2000)

WP348 John Caligari, *The Army's Capacity to Defend Australia Offshore: The Need for a Joint Approach* (2000)

WP349 He Kai, *Interpreting China-Indonesia Relations: 'Good-Neighbourliness', 'Mutual Trust' and 'All-Round Cooperation'* (2000)

WP350 Paul Dibb, *Strategic Trends in the Asia-Pacific Region* (2000)

WP351 Andrew Selth, *Burma's Order of Battle: An Interim Assessment* (2000)

WP352 Andrew Selth, *Landmines in Burma: The Military Dimension* (2000)

WP353 Desmond Ball and Euan Graham, *Japanese Airborne SIGINT Capabilities* (2000)

WP354 Bilveer Singh, *The Indonesian Military Business Complex: Origins, Course and Future* (2001)

WP355 R.S. Merrillees, *Professor A.D. Trendall and his Band of Classical Cryptographers* (2001)

WP356 Desmond Ball and Hazel Lang, *Factionalism and the Ethnic Insurgent Organisations in the Thailand-Burma Borderlands* (2001)

WP357 Ron Huisken, *ABM vs BMD: The Issue of Ballistic Missile Defence* (2001)

WP358 Greg Mills and Martin Edmonds, *South Africa's Defence Industry: A Template for Middle Powers?* (2001)

WP359 Desmond Ball, *The New Submarine Combat Information System and Australia's Emerging Information Warfare Architecture* (2001)

WP360 Desmond Ball, *Missile Defence: Trends, Concerns and Remedies* (2001)

WP361 Matthew N. Davies, *Indonesian Security Responses to Resurgent Papuan Separatism: An Open Source Intelligence Case Study* (2001)

WP362 Ron Huisken, *ANZUS: Life after 50. Alliance Management in the 21st Century* (2001)

WP363 Ron Huisken, *A Strategic Framework for Missile Defence* (2002)

WP364 Coral Bell, *The First War of the 21st Century: Asymmetric Hostilities and the Norms of Combat* (2002)

WP365 Paul Dibb, *The Utility and Limits of the International Coalition against Terrorism* (2002)

WP366 Ron Huisken, *QDR 2001: America's New Military Roadmap: Implications for Asia and Australia* (2002)

WP367 Andrew Tan, *Malaysia's Security Perspectives* (2002)

WP368 Ron Huisken, *Asia Pacific Security Taking Charge, Collectively* (2002)

WP369 Paul Dibb, *The War on Terror and Air Combat Power: A Word of Warning for Defence Planners* (2002)

WP370 John McFarlane, *Organised Crime and Terrorism in the Asia-Pacific Region: The Reality and the Response* (2002)

WP371 Clive Williams, *The Sydney Olympics — The Trouble-free Games* (2002)

WP372 Ron Huisken, *Iraq: America's Checks and Balances Prevail over Unilateralism* (2003)

WP373 Alan Dupont, *The Kopassus Dilemma: Should Australia Re-engage?* (2003)

WP374 Alan Dupont, *Transformation or Stagnation?: Rethinking Australia's Defence* (2003)

WP375 Brek Batley, *The Justifications for Jihad, War and Revolution in Islam* (2003)

WP376 Lieutenant-Colonel Rodger Shanahan, *Radical Islamist Groups in the Modern Age: A Case Study of Hizbullah* (2003)

WP377 Andrew Selth, *Burma's China Connection and the Indian Ocean Region* (2003)

WP378 David Connery, *Trash or Treasure? Knowledge Warfare and the Shape of Future War* (2003)

WP379 Christian Enemark, *Biological Weapons: An Overview of Threats and Responses* (2003)

WP380 Desmond Ball, *Security Trends in the Asia-Pacific Region: An Emerging Complex Arms Race* (2003)

WP381 Brice Pacey, *National Effects-Based Approach: A Policy Discussion Paper* (2003)

WP382 Desmond Ball, *China's Signals Intelligence (SIGINT) Satellite Programs* (2003)

WP383 Alan Stephens, *The End of Strategy: Effects-Based Operations* (2003)

WP384 David Bolton, *The Tyranny of Difference: Perceptions of Australian Defence Policy in Southeast Asia* (2003)

WP385 Ron Huisken, *The Threat of Terrorism and Regional Development* (2004)

WP386 Ron Huisken, *America and China: A Long Term Challenge for Statesmanship and Diplomacy* (2004)

WP387 Damien Kingsbury, *Australia's Renewal of Training Links with Kopassus* (2004)

WP388 Desmond Ball, *How the Tatmadaw Talks: The Burmese Army's Radio Systems* (2004)

WP389 Sandy Gordon, *Muslims, Terrorism and Rise of the Hindu Right in India* (2004)

WP390 Ron Huisken, *We Don't Want the Smoking Gun to be a Mushroom Cloud: Intelligence on Iraq's WMD* (2004)

WP391 Gavin Keating, *Opportunities and Obstacles: Future Australian and New Zealand Cooperation on Defence and Security Issues* (2004)

WP392 Brendan Taylor, *Instability in the US-ROK Alliance: Korean Drift or American Shift?* (2004)

WP393 Ron Huisken, *North Korea: Power Play or Buying Butter With Guns?* (2004)

WP394 Christopher Michaelsen, *The Use of Depleted Uranium Ammunition in Operation Iraqi Freedom: A War Crime?* (2005)

WP395 Alan Stephens, *A Threat-based Reassessment of Western Air Power* (2005)

WP396 Richard Brabin-Smith, *The Heartland of Australia's Defence Policies* (2005)

WP397 Pak Shun Ng, *From 'Poisonous Shrimp' to 'Porcupine': An Analysis of Singapore's Defence Posture Change in the early 1980s* (2005)

WP398 James D. Stratford, *Assisting the Solomon Islands: Implications for Regional Security and Intervention* (2005)

WP399 Ron Huisken, *Iraq: Why a Strategic Blunder Looked So Attractive* (2006)

WP400 Richard Brabin-Smith, *Australia's International Defence Relationships with the United States, Indonesia and New Zealand* (2006)

WP401 Desmond Ball, *The Probabilities of On the Beach: Assessing Armageddon Scenarios in the 21st Century* (2006)